REVISE AQA GCSE
Mathematics A
REVISION WORKBOOK
Higher

Series Consultant: Harry Smith   Author: Greg Byrd

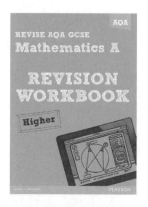

THE REVISE AQA SERIES
Available in print or online

Online editions for all titles in the Revise AQA series are available Spring 2013.

Presented on our ActiveLearn platform, you can view the full book and customise it by adding notes, comments and weblinks.

Print editions

Mathematics A Revision Workbook Higher   9781447941446

Mathematics A Revision Guide Higher   9781447941361

Online editions

Mathematics A Revision Workbook Higher   9781447941514

Mathematics A Revision Guide Higher   9781447941378

This Revision Workbook is designed to complement your classroom and home learning, and to help prepare you for the exam. It does not include all the content and skills needed for the complete course. It is designed to work in combination with Pearson's main AQA GCSE Mathematics 2010 Series.

To find out more visit:
www.pearsonschools.co.uk/aqagcsemathsrevision

Contents

- - - - - - - - - - - - - - - - - - - -

A small bit of small print

AQA publishes Sample Assessment Material and the Specification on its website. That is the official content and this book should be used in conjunction with it. The questions in this book have been written to help you practise every topic in the book. Remember: the real exam questions may not look like this.

Target grades
Target grades are quoted in this book for some of the questions. Students targeting this grade should be aiming to get most of the marks available. Students targeting a higher grade should be aiming to get all of the marks available.

Calculator skills 1

 ✓ **1** Use your calculator to work out $\dfrac{11.9 + 4.3}{7.5 - 3.7}$ $\dfrac{16.2}{3.8}$

(a) Write down your full calculator display.

Answer .... *(1 mark)*

(b) Write down the answer correct to 1 decimal place.

Answer .... *(1 mark)*

 ✓ **2** Use your calculator to work out $\dfrac{6.34^2}{2.65 \times 5.91}$

(a) Write down your full calculator display.

Answer .... *(1 mark)*

(b) Write down the answer, correct to 3 significant figures.

Answer .... *(1 mark)*

 ✓ **3** Put these approximations for π in order of size, starting with the smallest.

> To convert fractions to decimals you divide the numerator by the denominator. Write down all the digits from your calculator display.

Guided

$$\dfrac{22}{7}, \quad \dfrac{25}{8}, \quad \dfrac{54}{17}$$

$\dfrac{22}{7} = 3.14285714$ $\dfrac{25}{8} = 3.125$ $\dfrac{54}{17} =$ ....

Answer .... *(2 marks)*

 ✓ **4** Here are some fractions, percentages and decimals. Put them in order of size, starting with the smallest.

$$\dfrac{3}{8}, \quad 35\%, \quad 0.345, \quad \dfrac{9}{25}$$

....

....

Answer .... *(2 marks)*

✓ **5** A pair of trainers costs £65 before the sale.

How much money is saved in the sale?

....

Answer £ .... *(2 marks)*

✓ **6** On average, Manchester has 140 days of rain per year. A year has 365 days.
What percentage of the year does Manchester have rain?
Give your answer to 1 decimal place.

....

....

Answer ....% *(3 marks)*

Percentage change 1

D ✓ **1** A car company increased the prices of its cars by 3.8%.
Before the increase, a car cost £12 950.
Work out the price after the increase.

Guided

> This is the amount the price has increased by. Now add it to the original cost of the car.

$$\frac{3.8}{100} \times 12\,950 = \dots\dots\dots\dots\dots$$

.... + 12 950 = ...   Answer £ ....   *(3 marks)*

> Write down all the digits off your calculator display, then give to a sensible degree of accuracy at the **end** of the question. As this is money, it would be sensible to round to 2 decimal places.

D ✓ **2** Last year the village fete made £838. This year it made 15% less.
Work out how much the village fete made this year. Give your answer to the nearest pound.

....

....

....

Answer £ ....   *(4 marks)*

D ✓ **3** A theatre ticket costs £29.50.
The price increases by 8%.
Pete has £100.
Can Pete buy three tickets at the new price? You **must** show your working.

....

....

....

....   *(4 marks)*

C ✓ **4** Pam bought a painting for £125 and
sold it for £155.
Calculate Pam's percentage profit.

> Remember that percentage profit and percentage increase are calculated in the same way.

....

....

....

Answer ....%   *(3 marks)*

C ✓ **5** Patrick estimates the length of a desk to be 95 cm. The actual length could be 10% higher or
10% lower. Work out the maximum and minimum possible lengths of the desk.

....

....

....

....

Maximum .... *cm*   *Minimum* .... *cm*   *(4 marks)*

Reverse percentages and compound interest

C ✓

1 Paavo invested £5000.

The interest rate is set at 4% compound interest per annum for three years.

How much is Paavo's investment worth, to the nearest pound, at the end of the three years?

> Per annum means per year.

Guided

Year	Balance (£)	Interest earned (£)
1	5000	$\frac{4}{100} \times 5000 = 200$
2	5200	
3		

Balance after 3 years = £ ....

> For year 2, you need to find 4% of £5200.

Answer .... (3 marks)

B

2 In a sale, all prices are reduced by 20%.

The sale price of a jacket is £92.

Work out the normal price of the jacket.

Guided

EXAM ALERT

> Students have struggled with exam questions similar to this – **be prepared!**

$100\% - 20\% = ...\ \%$

$92 \div \dfrac{.........}{100} = $

> Check your answer by reducing the normal price by 20%. You should get £92 as your result.

Answer £ .... (3 marks)

B

3 A large water container contains 2000 litres. The container has a hole in and the water leaks out at a rate of 7% an hour.

After how many hours does the container hold less than $\frac{3}{4}$ of its original amount?

> This is a compound percentage question. Think of it as repeated percentage loss.

....

....

....

....

Answer ....4......... hours   (3 marks)

A ✓

4 A woman's salary increased by 8% in one year and reduced by 8% in the next year.

Is her final salary greater or less than her original salary, and by how many per cent?

....

....

....

....

....

....

Answer .... %   (3 marks)

Ratio

D ✓ **1** Write the ratio 4 : 10 in its simplest form.

....

Answer ....2 : 5............ *(1 mark)*

D ✓ **2** Write the ratio 6 : 12 : 24 in its simplest form.

....

Answer ....1 : 2 : 4............ *(1 mark)*

D ✓ **3** Write the ratio 8 : 18 in the form 1 : n.

............2.25............

Answer ....1 : 2.25............ *(1 mark)*

D **4** Write two ratios that are equivalent to 2 : 5. | Your answers must both be able to be simplified to 2 : 5.

............4 : 10............

............8 : 20............

Answer ....4 : 10............ and ....8 : 20............ *(2 marks)*

D **5** Share £48 in the ratio 1 : 5.

> **Guided**

Number of parts = 1 + 5 = ...6............

One part is worth = £48 ÷ ....6............

= £ ....8............

The ratio has a total of 6 parts. Next, work out what each part is worth.

1 × £8 = ...8............ and 5 × £8 = ...40............

Answer £............8............ and £............40............ *(2 marks)*

C ✓ **6** The ratio of boys to girls at a party is 4 : 7. There are 24 boys at the party. How many girls are at the party?

> **Guided**

24 ÷ 4 = ...6............

7 × ....6............ = ...42............

Work out what each part of the ratio is worth.

Girls make up 7 parts of the ratio, so multiply the value of one part by 7.

Answer ....35............ *(2 marks)*

C ✓ **7** Richie and Amy eat a chocolate cake in the ratio 3 : 2. Richie eats 285 g of the cake. What weight of chocolate cake does Amy eat?

............95   3)285............ ....2 × 95 = 190............

............3)285............

....

Answer ....190............g *(3 marks)*

Standard form

B

1 Write 35 000 000 in standard form.

> Guided

3.5×10^7.

> The first number must be greater than or equal to 1 and less than 10.

Answer .... 3.5×10^7 .... *(1 mark)*

B

2 Write 6.01×10^5 as an ordinary number.

> Guided

6.01

> Remember that $\times 10^5$ means $\times 10 \times 10 \times 10 \times 10 \times 10$. Do five jumps, and write in some extra zeros!

Answer .... 601000 .... *(1 mark)*

B

3 Use your calculator to find the square root of 6.55×10^{11}.
Give your answer in standard form correct to 2 decimal places.

$6.55 \times 10^{11} = 655000000000.0$

Answer .... $809 \; \cancel{370.70}$ .... *(2 marks)*
$\times 10^5$

B

4 Astronomers measure distances in space using AU (Astronomical Units).
1 AU is about 150 000 000 km.
The distance from our Sun to the next closest star, Proxima Centauri is about 271 000 AU.
How many kilometres is our Sun from Proxima Centauri?
Give your answer in standard form.

$271 \times 150,000$ $13550\,000$ $1000 \times 1\,000\,000 = 1\,bn$
271 $+ 27100,000$
$\times 150\,000$ $4065\,000\,000,000$
$13550\,000$
$27100\,0\,00$ Answer .... 4.065×10^{13} .... km   *(3 marks)*

B

5 A, B and C are standard form numbers: $A = 5.5 \times 10^9$, $B = 3.3 \times 10^{-8}$, $C = 1.5 \times 10^{11}$.
Calculate the following, giving all answers in standard form:

(a) $A + C = $ .... 1.555×10^{11}

Answer .... 1.555×10^{11} .... *(1 mark)*

(b) $B^2 = $ .... 1.089×10^{-15}

Answer .... 1.089×10^{-15} .... *(1 mark)*

(c) $\dfrac{AB}{C} = $ .... $1.815 \times 10^2 \div \cancel{1.21 \times 10^9} \; 1.5 \times 10^{11}$

Answer .... $\cancel{1.815 \times 10^2}$ .... *(1 mark)*
1.21×10^9

B

6 The mass of one atom of carbon is 1.994×10^{-23} kg.
The mass of one atom of oxygen is 2.66×10^{-23} kg.
One molecule of carbon dioxide gas has **one** atom of carbon and **two** atoms of oxygen.
Work out the total mass of one molecule of carbon dioxide.

$1.994 \cancel{\times 10^{-23}}$
2.66
$+ 2.66$
$\overline{7.314}$ Answer .... 7.314×10^{-23} .... kg   *(2 marks)*

Upper and lower bounds

A

1 The liquid contents of two containers are given.
The smaller container is given to the nearest millilitre.
The larger container is given to 2 significant figures.

1500 ml

15 ml

> The greatest possible amount means you must add the upper bounds of both containers.

(a) Work out the greatest possible amount of liquid in the two containers.

Upper bound of small container

= 15 + 0.5 = ...15.5............ ml

Upper bound of big container

= 1500 + 50 = ...1550............ ml

> 1500 ml has been rounded to 2 significant figures. That is to the nearest 100, so add 50 to find the upper bound.

Greatest amount of liquid

= ....15.5............ + ...1550............ = ...1565.5............

Answer ....1565.5............ ml   *(2 marks)*

(b) Work out the greatest possible difference between the contents of the two containers.

> The greatest difference is found by subtracting the lower bound of the small container from the upper bound of the big container.

$\begin{array}{r} 1550.00 \\ - \quad 14.50 \\ \hline 1535.50 \end{array}$

Answer ....1535.5............ ml   *(2 marks)*

A

EXAM ALERT

2 Population density is the average number of people per square kilometre.
The population of Scotland is 5 200 000 to the nearest hundred thousand.
The area of Scotland is 78 800 km², correct to 3 significant figures.
Calculate the minimum population density in Scotland.
You **must** show your working.

> Students have struggled with exam questions similar to this – **be prepared!**

a = 5 150 000

b = 78 750

a ÷ b = c

Answer ....65............ people per square km   *(2 marks)*

A

3 Sound travels at 340 m/s (metres per second) to the nearest 10 m/s.
Liam sees a footballer kick a football and hears the kick 0.8 seconds later, correct to 1 decimal place.
What is the furthest the footballer could be away from Liam?

a = 345

b = 0.85

Answer ....293.25............ m   *(2 marks)*

The Data Handling Cycle

Guided

1 Dirk wants to test this hypothesis:

> It is cheaper to buy an iPod on the internet than in a high street store.

Use the Data Handling Cycle to write a plan for Dirk.

> Describe how you will collect data, and give at least two different statistics you could calculate. Explain how you will compare the cost of shopping online and on the high street.

Collect data by ....

Calculate ....

Compare ....

.... and write a conclusion.   *(3 marks)*

This is the Data Handling Cycle:

Specify the problem and plan an investigation

→ Collect data using a survey or experimenting

↓ Analyse and present your data using statistics, graphs and chart

← Interpret your results and make conclusions

---

2 Shilpa sees this poster in her school canteen.
Use the Data Handling Cycle to write a plan for Shilpa to decide whether this is true for her year group.

> **HEALTH FOOD FRIDAYS!**
> The number of students who choose the healthy option increases by 20% on Fridays.
> **WHY NOT JOIN THE CROWD AND GO LEAN AND GREEN THIS FRIDAY?**

....

....

....

....

....

....   *(3 marks)*

---

3 Danielle wants to test this hypothesis.

> Girls spend more time completing homework than boys.

Use the Data Handling Cycle to write a plan for Danielle.

> You have been given the hypothesis, so you do not have to include it in your answer.

....

....

....

....

....   *(3 marks)*

Collecting data

D **1** Match each data collection method to **one** situation.

Guided

Watch how people react to a loud noise	Data Logging
Obtain opinions on a new perfume	Observation
Collect data on reaction times of students	Experiment
Record number of people through a turnstile	Questionnaire

> If you are watching for a reaction, you are **observing**.

D **2** Is shoe size discrete or continuous data?
Tick a box.

> You can *only* have sizes such as 3, $3\frac{1}{2}$, 4, $4\frac{1}{2}$, etc.

Discrete ☑   Continuous ☐   *(1 mark)*

Give a reason for your answer.
.... There are set shoe sizes ....
.... *(2 marks)*

D **3** Is the length of a shoe discrete or continuous data?
Tick a box.

Discrete ☐   Continuous ☑   *(1 mark)*

Give a reason for your answer.
.... There are not set lengths of feet ....
.... *(2 marks)*

D **4** Cathy is conducting research on apple trees in her orchard.
Use the words qualitative, quantitative, discrete and continuous to describe the sort of data she collects.

Guided

Data	Type of data
Number of trees in orchard	quantitative, discrete
Species of apple tree	quantitative
Diameter of tree trunk	continuous,
Number of apples per tree	quantitative discrete
Weight of apples per tree	continuous,
Colour of apple	qualitative

> A number is quantitative data, and you can only have a whole number of trees.

(3 marks)

D **5** Cherry collected data from the internet about the weather in Spain last summer.
She said, 'I collected the data myself, so it is primary data'.
Cherry is wrong. Explain why Cherry is wrong.
.... She did not collect it, the weather people did. ....
.... It was Secondary data ....
.... *(2 marks)*

Surveys

EXAM ALERT

1 Sammi claims that GCSE Geography students are more likely than other students to know where Newcastle is on a map of the UK. Design a data collection sheet for Sammi to investigate her claim.

> Students have struggled with exam questions similar to this – **be prepared!**

> A data collection sheet and an observation sheet are the same thing.

> Sammi needs to find out whether a student is studying GCSE Geography or not and whether they can point to the location of Newcastle on a map of the UK or not.

(2 marks)

2 Simon is writing a questionnaire on the amount of television watched by students on week nights. This is one of his questions:

> How much time do you spend watching TV on Mondays?
>
> Not much ☐   About average ☐   Too much ☐

(a) Give one reason why Simon's question might not be a good one.

No given times, not enough options

(1 mark)

(b) Using Simon's question, design a better response section.

> Write at least 4 different response boxes. Choose numbers so that every person can tick exactly one box.

☐ 0 hours   ☐ 1-2   ☐ 3-4   ☐ 5-6   ☐ 7-8

(1 mark)

3 Stan wants to survey students in his Maths class to find out how long it takes them to get to school in the morning. Write a suitable question. Include a response section.

....

....

....

(2 marks)

Two-way tables

1 100 students study one of the three GCSE subject options listed in the table below.
The two-way table shows some information about these students.

Guided

> Male / Total is 100 − 45 = 55.

	Geography	History	Sports Studies	Total
Male	12	18	25	55
Female	18	10	17	45
Total	30	28	42	100

> Geography / Total is 12 + 18 = 30.

> Male / Geography is 55 − 25 − 18 = 12.

(a) Complete the two-way table.   *(4 marks)*

(b) How many female students study History?   Answer ....10.........   *(1 mark)*

(c) What percentage of the students study Sports Studies?

....

                        Answer ....42.........%   *(1 mark)*

2 The two-way table gives the lunchtime activities of 75 students.

	Sports	Library	Chatting	Total
Male	8	12	13	33
Female	21	16	5	42
Total	29	28	18	75

(a) Complete the two-way table.   *(4 marks)*

(b) How many of the females spent their lunchtime chatting?

                              Answer ....5.........   *(1 mark)*

(c) What fraction of the male students spent their lunchtime in the library?

....

                              Answer ....$\frac{12}{33}$.........   *(1 mark)*

3 Toni is collecting some information as to whether the students in three classes thought that their
English test was a fair test or not.
There are 90 students in total.
There are 6 more male than female students.
Two thirds of the students thought the test was not fair.
One quarter of the males thought the test was fair.
Use this information to complete the two-way table below.

	Male	Female	Total
Thought test was fair	12	18	30
Thought test was not fair	36	24	60
Total	48	42	90

....

....   *(4 marks)*

Sampling

1 The local council is doing a survey to find opinions as to whether to build a new running track or a basketball court at the local sports centre.

Guided

Give **two** reasons why asking 15 runners would **not** make a good sample.

> For the second reason, explain why only asking runners would not give a meaningful result.

1   15 people would not be enough to give a fair sample.

2   ......Bias...

......didn't ask enough of a variety of people..........................

(2 marks)

2 Give two conditions you would need to satisfy if you wanted a random sample of 30 students from your school?

1   .....50% male & 50% female.....................................

.....Only my school...

2   .....Different ages..

.....Equal opportunity............................. *(2 marks)*

3 What is meant by a stratified sample?

....

....

.... *(2 marks)*

EXAM ALERT

4 A university has arts students and sciences students as shown.

> Students have struggled with exam questions similar to this – **be prepared!**

Type	Arts students	Sciences students
Number of students	2850	1780

Siobhan carries out a survey of the students.
She uses a sample of 120 students, stratified by type.
Work out the number of arts students in her sample.

....

....

....

Answer.......74...... *(2 marks)*

5 Sally questions a sample of farm owners in one county, stratified by gender.
At the beginning of 2012 there were 216 female farm owners.
This was 29.2% of the total number of farm owners in that county.
Complete the table to show the number of male and female farm owners in a sample of 50 farm owners.

Male	35
Female	15

292-216

....

....

Answer .... *(2 marks)*

Mean, median and mode

1   The table shows the lengths of 100 four-week-old tadpoles.

Length, l (mm)	Frequency
$8 \leqslant l < 10$	42
$10 \leqslant l < 12$	58

Write down the modal class.

The modal class is the class with the highest frequency.

Answer ....10⩽l<12.... *(1 mark)*

2   The table shows the lengths of 100 ten-week-old tadpoles.

Length, l (mm)	Frequency
$8 \leqslant l < 10$	4
$10 \leqslant l < 12$	15
$12 \leqslant l < 14$	30
$14 \leqslant l < 16$	29
$16 \leqslant l < 18$	22

The median is the central, or middle, value. It will be halfway between the 50th and 51st values.

Write down the class interval which contains the median.

Answer ....14<l<16.... *(1 mark)*

3   (a) Write down a possible set of 5 numbers that have a mode of 5 and a median of 7.

Answer .....5........ ......5........ ......7........ ......8........ ......9........ *(1 mark)*

(b) Write down a different possible set of 5 numbers that have a mode of 5 and a median of 7.

Answer .....5........ ......5........ ......7........ ......10........ ......11........ *(1 mark)*

4   Ten numbers have a median of 7. The numbers are written in order of size.

1   2   2   4   5   〔7〕 9   10   11   12   15

> **Guided**

Work out the missing value.

$$\frac{5 + \boxed{9}}{2} = 7$$

Median is halfway between the 5th and 6th values.

Answer ....9........ *(2 marks)*

5   Martin wants to find the average of 7 numbers: 1, 3, 3, 3, 4, 6, 28
He decides to find the mean.
Why might this not be the best average to find? Give a reason for your answer.

..........28 is an outlier, it will ruin the average..........

.... *(2 marks)*

6   The mean, mode and median of six numbers are all 50% of the largest number.
Work out a possible set of six numbers.

....

....

....

Answer .....0...... ......2........ ......2........ ......2........ ......2........ ......4........ *(3 marks)*

Frequency table averages

1   Faye collects data about the number of televisions in different families.

Number of televisions	Number of families	$f \times x$
0	1	$0 \times 1 = 0$
1	0	$1 \times 0 = 0$
2	15	$2 \times 15 = 30$
3	8	$3 \times 8 = 24$
4	4	$4 \times 4 = 16$
Total	28	70

> f is the frequency, or the number of families and x is the number of TVs per family.

> Complete the next two rows in the $f \times x$ column. You calculate the mean by finding the total of the $f \times x$ column and then dividing by the total number of families.

Work out the mean number of televisions in each family.

$$28\overline{)70} \quad \begin{array}{r} 2.5 \\ 28\overline{)140} \\ -56 \\ \hline 14 \end{array}$$

Answer .... 2.5 ....   *(4 marks)*

2   Fred works part-time at a local amusement arcade.
He is paid £7.40 an hour.
The busier the arcade, the more hours he works.
Fred records the number of hours he works per week.

> Students have struggled with exam questions similar to this – **be prepared!**

 EXAM ALERT

Number of hours, h	Frequency, f	Midpoint of the number of hours, x	$f \times x$
$10 \leqslant h < 12$	6	11	66
$12 \leqslant h < 14$	12	13	156
$14 \leqslant h < 16$	20	15	300
$16 \leqslant h < 18$	5	17	85
Total	43	56	607

Work out an estimate of his mean weekly pay from the arcade.

Answer £ ...... 104.46 ......   *(4 marks)*

Interquartile range

B

Guided

1 Isobel has measured the tail length, in cm, of 11 stingrays.

| 35 | 21 | 34 | 40 | 26 | 29 | 29 | 30 | 24 | 33 | 30 |

To find the interquartile range (IQR) first put your data in order.

Work out the interquartile range of the stingray tail length.

21, 24, 26, 29, 29, 30, 30, 33, 34, 35, 40

The lower quartile = $\frac{n+1}{4}$ = $\frac{11+1}{4}$ = $\frac{12}{4}$ = 3rd piece of data.

$n = 11$, as there are 11 pieces of data.

LQ = 26

The upper quartile = $\frac{3(n+1)}{4}$ = $\frac{3(11+1)}{4}$ = $\frac{36}{4}$ = 9th piece of data. UQ = 9th = 34

The IQR = LQ − UQ = 26 − 34 = 8

Answer 8 cm   *(3 marks)*

B

2 15 runners each ran 200 metres. The time taken was recorded to the nearest second.

37 25 28 33 42 40 33 37

36 38 22 39 26 42 31

Always start by putting the data in order.

Work out the interquartile range of this data.

$\frac{16}{4}$ = $\frac{4}{1}$ = 28

$\frac{48}{4}$ = $\frac{12}{4}$ = 39

Answer 11 s   *(3 marks)*

B

3 Some athletes each threw a javelin. The distance was recorded to the nearest metre.
The stem-and-leaf diagram shows this information.

Athlete's distance

2	3	8	9								
3	0	2	2	2	4	5	5	7	8	8	8	9
4	0	0	2	3	9							
5	1	4										
6	6											

Key: 2 | 3 = 23 metres

Work out the median and the interquartile range of this data.

$\frac{n+1}{4}$ $\frac{3(n+1)}{4}$ $\frac{24-6}{4-1}$ $\frac{72}{4} - \frac{18}{1}$ 32−42

Median 38 m   Interquartile range 10 m   *(3 marks)*

Scatter graphs

1   The table shows the costs in pence and numbers of pages of 10 sudoku puzzle books.
All the values have been given to the nearest 10.

Guided

Cost (pence)	60	130	240	260	80	120	130	80	200	60
Number of pages	60	100	200	230	70	100	90	60	140	50

(a) Complete the scatter diagram.

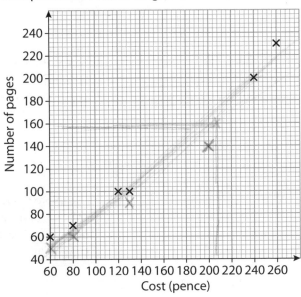

The first six points have been plotted for you.

(2 marks)

(b) Draw a line of best fit for this data.   *(1 mark)*

(c) What can you say about the correlation of this data?

.... *positive* ....

All you need to say is **one** of: positive correlation, negative correlation or no correlation.

(1 mark)

(d) Describe the connection between the cost of a sudoku puzzle book and the number of pages it contains.

.... *the more pages, the higher the cost* .... *(1 mark)*

(e) Use your line of best fit to estimate the cost of a sudoku puzzle book which contains 160 pages.

Answer .... *215* .... pence   *(1 mark)*

2   The table shows the marks scored by 10 students in an English test and a Chemistry test.

Student	A	B	C	D	E	F	G	H	I	J
English	78	66	58	54	72	44	34	50	24	30
Chemistry	40	38	44	56	36	58	66	68	68	76

(a) Use the axes opposite to plot a scatter graph for this data.   *(2 marks)*

(b) Draw a line of best fit for this data.   *(1 mark)*

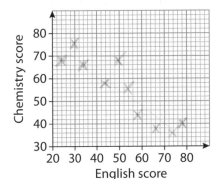

Frequency polygons

C

Guided

EXAM ALERT

1   A company employs 20 staff.
The table shows information about their salary.

Students have struggled with exam
questions similar to this – **be prepared!**

Salary, s (£1000)	Frequency	Midpoint
$10 < s \leqslant 20$	14	15
$20 < s \leqslant 30$	4	25
$30 < s \leqslant 40$	1	35
$40 < s \leqslant 50$	1	45

(a) Show this information on a frequency polygon.

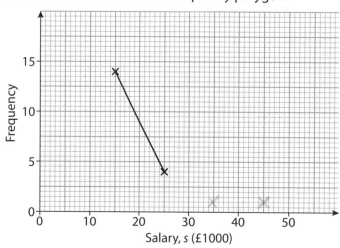

The midpoint of $10 < s \leqslant 20$
is found by adding 10 and 20,
and then dividing by 2.

Complete the midpoint
column, plot the last two
points and join the points with
straight lines.

(2 marks)

(b) Work out an estimate of the mean salary.

$$\text{Mean} \approx \frac{\text{sum of (frequency} \times \text{midpoint)}}{\text{total frequency}}$$

$$= \frac{(14 \times 15\,000) + (4 \times 25\,000) + (1 \times 35\,000) + (1 \times 45\,000)}{14 + 4 + 1 + 1}$$

$$= \frac{210\,000 + 100,000 + 35\,000 + 45\,000}{20}$$

$$\begin{array}{l} 19500 \\ 12750 \end{array}$$

$$= \underline{\hspace{6cm}}$$

Answer £ ....19500.......   *(3 marks)*

Histograms

1 The table shows information about the mass, in kg, of the suitcases on a plane.

Mass, m (kg)	Frequency	Frequency density
$5 \leqslant m < 15$	58	5.8
$15 \leqslant m < 18$	14	4.7
$18 \leqslant m < 20$	11	5.5
$20 \leqslant m < 21$	5	5
$21 \leqslant m < 31$	22	2.2

110

$$\text{Frequency density} = \frac{\text{Frequency}}{\text{Class width}}$$

(a) On the grid below, show the data on a histogram. *(3 marks)*

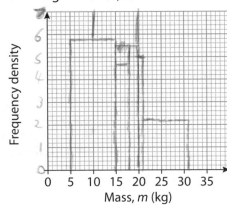

Histograms have frequency density on the y-axis.

Work out all of the frequency densities before choosing a scale for the y-axis of the graph.

(b) Estimate the percentage of suitcases that have a mass of between 10 kg and 20 kg.

Draw a line up from 10 kg and from 20 kg. Work out the area under the histogram between your lines to estimate the number of suitcases that have a mass of between 10 kg and 20 kg.

$$\frac{58 - 29 + 14 + 11}{2} = 54$$

Answer ....49.1........%   *(2 marks)*

2 The speed of 485 cars is shown in the following table and histogram.

Speed (mph)	Frequency	Frequency density
$15 \leqslant s < 20$	90	18
$20 \leqslant s < 25$	130	26
$25 \leqslant s < 29$	120	30
$29 \leqslant s < 36$	105	15
$36 \leqslant s < 40$	40	10

Histogram of speed of cars

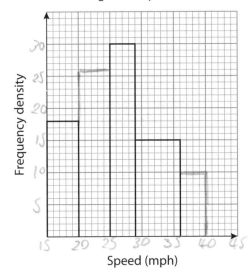

Complete the histogram and table.

....

.... *(4 marks)*

Probability 1

1 Peter has a biased dice, numbered 1 to 6.
The table shows the probability of rolling
some of the numbers on the dice.

Number	Probability
1	0.1
2	0.3
3	0.2

(a) What is the probability of **not** rolling a 1?

P(not 1) = 1 − P(1) = 1 − ....   Answer ....0.9........ *(1 mark)*

(b) What is the probability of rolling a 2 or a 3?

P(2) + P(3) =
0.2 + 0.3 = 0.5

Answer ....0.2 + 0.3 = 0.5........ *(2 marks)*

(c) What is the probability of rolling a 4, 5 or 6?

.1 + .2 + .3 = .6

Answer ....P(1) + P(2) + P(3) = ...6 − 1 = .4 *(1 mark)*

2 A four-sided dice, numbered 1 to 4, and a coin are thrown at the same time.

(a) Draw a sample space diagram to show all possible outcomes.

> To start, you need to draw a grid, with the numbers of the dice on the top, and the options of the coin down the left hand side.

	1	2	3	4
H	.125	.125	.125	.125
T	.125	.125	.125	.125

(2 marks)

(b) What is the probability of getting a head and a 4?

Answer ....0.125........ *(1 mark)*

3 The probability of X occurring is 0.2. The probability of Y occurring is 0.3.

(a) When X and Y are **mutually exclusive** events, work out the probability of X **or** Y occurring.

Answer ....0.5........ *(1 mark)*

(b) When X and Y are **independent** events, work out the probability of X **and** Y occurring.

Answer ....0.06........ *(2 marks)*

C **4** In a game, players try to win a coloured counter.
There are five possible colours.
The table shows the probability of winning each colour.

Colour of Counter	Probability
Red 1	0.07
White 2	0.23
Blue 3	0.05
Green 4	0.14
Black 5	0.21

Pierre plays the game 3 times. What is the probability that he does not win a counter?

1 − (P(1) + P(2) + P(3) + P(4) + P(5)) = 0.3   0.3 × 0.3 × 0.3 = 0.027

Answer ....0.027........ *(2 marks)*

Probability 2

 1 Philip has a biased six-sided dice.
The probabilities of it landing on each number are given in the table.
Philip rolls the dice 100 times.
Work out an estimate for the number of times the dice will land on a 3.

Number	1	2	3	4	5	6
Probability	0.1	0.2	0.3	0.2	0.1	0.1

Answer ...... 100 × 0.3 = 30 ...... *(2 marks)*

  2 A teacher measured the hand spans of the students in her class.
She recorded the results in a frequency table.
The teacher chooses a student at random.

Hand span, h (cm)	Frequency
$12 \leq h < 14$	4
$14 \leq h < 16$	6
$16 \leq h < 18$	13
$18 \leq h < 20$	7

(a) Estimate the probability that the student has a hand span of 18 cm or more.

Total number of students = 4 + 6 + 13 + 7 = ...30...

$P(h \geq 18) = \dfrac{7}{30} = $ ......

Answer ...... 0.2333 ...... *(2 marks)*

(b) Estimate the probability that the student has a hand span of less than 16 cm.

$P(h<16) = \dfrac{10}{30} = \dfrac{1}{3} = 0.333$

Answer ...... 0.333 ...... *(2 marks)*

3 A geologist weighed a sample of 45 pebbles from a beach.
He recorded the results in a frequency table.

Weight, w (g)	Frequency
$40 \leq h < 50$	15
$50 \leq h < 60$	12
$60 \leq h < 70$	10
$70 \leq h < 80$	8

(a) The geologist picks up two more pebbles from the beach at random.
Estimate the probability that both pebbles weigh less than 50 g.

Assume the probabilities are the same for both picks.

$P(Pebble 1 = w<50) \times P(Pebble 2 = w<50)$

$P(w<50) = \dfrac{15}{45} = \dfrac{1}{3}$ $\dfrac{1}{3} \times \dfrac{1}{3} = \dfrac{1}{9} = 0.111$

Answer ...... 0.111 ...... *(3 marks)*

(b) Comment on the accuracy of your estimate.

Accuracy suffers because of small sample size.

Answer ...... *(1 mark)*

 4 A software company has two different DVD burners, A and B.
The probability that DVD burner A produces a faulty DVD is 0.1.
The probability that DVD burner B produces a faulty DVD is 0.05.
On one day the company burns 850 DVDs with burner A, and 1500 DVDs with burner B.

Estimate the total number of faulty DVDs produced on that day.

850 × 0.1 = 85   85 + 75 = 160
1500 × 0.05 = 75

Answer ...... 160 ...... *(4 marks)*

Tree diagrams

Guided

1 A bag contains 3 red counters and 5 green counters.
Theresa takes a counter at random from the bag, notes its colour and **replaces** it. She then takes another counter at random from the bag.

(a) Complete the tree diagram:

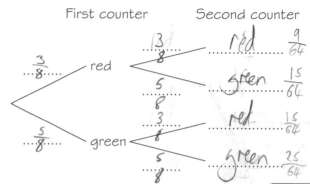

First counter   Second counter

red $\frac{3}{8}$ → red $\frac{9}{64}$

$\frac{3}{8}$ red → green $\frac{15}{64}$

green $\frac{5}{8}$ → red $\frac{15}{64}$

$\frac{5}{8}$ green → green $\frac{25}{64}$

(2 marks)

(b) Work out the probability that the two counters are both red.

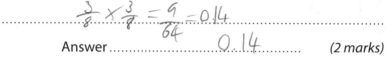
$$\frac{3}{8} \times \frac{3}{8} = \frac{9}{64} = 0.14$$

Answer .... 0.14 .... *(2 marks)*

> Multiply the probability that the first counter is red by the probability that the second counter is red.

(c) Work out the probability that the two counters are of different colours.

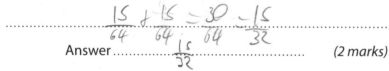
$$\frac{15}{64} + \frac{15}{64} = \frac{30}{64} = \frac{15}{32}$$

Answer .... $\frac{15}{32}$ .... *(2 marks)*

> There are two separate outcomes: red then green, and green then red.

A* **2** Terry has a drawer with 6 black socks and 5 red socks.
He gets dressed in the dark, taking two socks from the drawer at random.
Work out the probability that both the socks are the same colour.

> When Terry takes the first sock from the draw he does **not** replace it, so there is one less sock to choose from when he takes his second sock.

$\frac{6}{11}$ black $\frac{5}{10}$ black $\frac{30}{110}$

$\frac{5}{10}$ red $\frac{30}{110}$ $30+20 = \frac{50}{110} = \frac{5}{11}$

$\frac{5}{11}$ red $\frac{6}{10}$ black $\frac{30}{110}$

$\frac{4}{10}$ red $\frac{20}{110}$

Answer .... $\frac{5}{11}$ .... *(4 marks)*

A* **3** Tom and Sam have been practising percentage questions for their Maths exam.
The probability that Tom gets a question correct is 0.8.
The probability that Sam gets a question correct is 0.6.
They both attempt another question.
Work out the probability that **at least one** of them gets it correct.

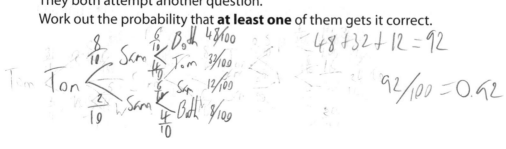

$\frac{8}{11}$ Tom $\frac{6}{10}$ Sam Both 48/100 $48+32+12 = 92$

Tom $\frac{6}{10}$ Sam Tom 32/100

$\frac{2}{10}$ Sam $\frac{6}{10}$ Sam Sam 12/100 $92/100 = 0.92$

$\frac{4}{10}$ Both 8/100

Answer .... 0.92 .... *(4 marks)*

Cumulative frequency

1 The table shows a summary of the time it took 80 students to complete a puzzle.

(a) Draw a cumulative frequency graph for the data.

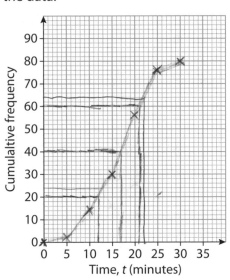

Time, t (minutes)	Frequency	Cumulative frequency
$0 < t \leq 5$	2	2
$5 < t \leq 10$	12	14
$10 < t \leq 15$	16	30
$15 < t \leq 20$	26	56
$20 < t \leq 25$	20	76
$25 < t \leq 30$	4	80

> A cumulative frequency graph has cumulative frequency up the y-axis. The second entry is 14 because $2 + 12 = 14$. The third entry is $14 + 16 = 30$.

> Plot the first point at (0, 0). Plot the remaining points at the **upper** end of each class interval. The second point is at (2, 5).

(b) Estimate the median for this data.

Answer .... 17 minutes .... *(1 mark)*

(c) Work out the interquartile range for this data.

$\frac{80}{4} = 20 = LQ$ $\frac{3 \times 80}{4} - 60 = UQ$ $LQ - UQ =$ $12 - 21 = 9$

Answer .... 9 minutes .... *(2 marks)*

(d) Chrissie says, 'I estimate that 30% of students will take more than 20 minutes to complete the puzzle.'
Is her estimate correct?
You **must** show your working.

$t \geq 20 = 24$ $\frac{24 \times 100}{80} = 30\%$

Answer .... Yes .... *(2 marks)*

(e) The school offered a prize for completing the puzzle quickly.

Finish in | 22 | minutes or under to win a prize.

80% of students won a prize. Use your cumulative frequency graph to estimate the missing number.

Answer .... 22 minutes .... *(2 marks)*

Box plots

1 Mrs McKay's students take an exam. The cumulative frequency graph shows the number of questions the students answered correctly.

> **Guided**

Before you work out the quartiles and median for the box plot you must find the total number of students – the total is **not** 50.

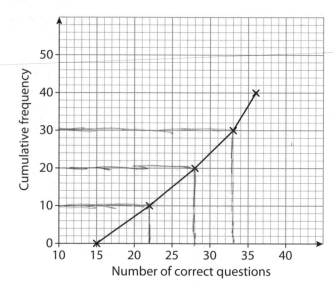

(a) Draw a box plot for Mrs McKay's students.

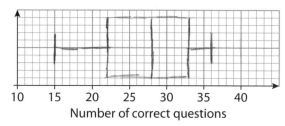

Number of correct questions

(5 marks)

19 of Mr Brown's students take the same exam.
The stem-and-leaf diagram shows the number of questions the students answered correctly.

```
1 | 8   9   9
2 | 0   0   1   3   5   6   6   8   8
3 | 0   1   2   2   2   3   4
```

Key: 1 | 8 represents 18 correct answers.

(b) Draw a box plot for Mr Brown's students.

$$\frac{n + 1}{2} = \frac{19 + 1}{2} = 10 \quad \text{Median = 10th value = } \underline{26}$$

$$\frac{n + 1}{4} = \frac{19 + 1}{4} = 5 \quad \text{Lower quartile = 5th value = } \underline{20}$$

$$\frac{3(n+1)}{4} = \frac{60}{4} = 15 \quad UQ = 32$$

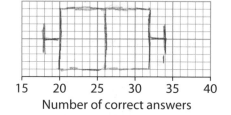

Number of correct answers

(5 marks)

(c) One of Mr Brown's students was ill.
Estimate the probability that the student would have got 32 or more marks.

Answer ....0.25 or 25%.......... *(1 mark)*

Comparing data

B **1** A company records how long its customers wait on hold when they call its customer service department. This box plot shows the results for a Tuesday morning.

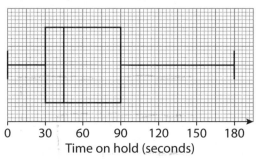

Time on hold (seconds)

(a) What is the interquartile range of time spent on hold?

....

Answer .... 60 .... seconds   *(1 mark)*

This table shows the company's results for a Saturday morning.

Minimum	Lower quartile	Median	Upper quartile	Maximum
20	35	40	120	180

(b) Compare the lengths of time customers spent on hold on Tuesday and on Saturday.

> You should compare an average, such as the median, and a measure of spread, such as the range or the interquartile range. You can use your answer to part **a**.

Tuesday Median = 45 seconds
IQ = 60 seconds
Saturday Median = 40 seconds   IQ = 85 seconds   *(1 mark)*

B **2** The box plots show information about the mass, in grams, of oranges picked at two different farms.

⟩Guided⟩

Compare the masses of the oranges picked at the two farms.

> Work out the median and interquartile range for each farm.

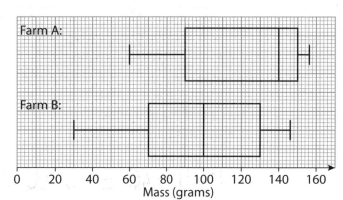

Farm A:

Farm B:

Mass (grams)

	Median (grams)	Interquartile range (grams)
Farm A	140	60
Farm B	100	60

....
Farm A has larger oranges. They win on minimum, maximum,
lower and upper quartiles and the median average.   *(3 marks)*
They only equal on the IQ.

23

Problem-solving practice 1

D 1 Philippe makes 150 models of Olympic athletes.
It costs £2.75 to make each model.
He keeps one-fifth of the models to give to friends and family.
He sells 80% of the rest for the full price of £9.
He sells the rest at half price.
How much profit does Philippe make?

$150 \times 2.75 = 412.5$ $330 \times 0.8 = 864$ 9720
$150 \times 0.8 = 120$ $120 - 96 = 24 \times 4.5 = 108$ $+108$ -412.5
$(120 \times 0.8) \times 9 = 864$ 9729 559.5
9728 559.50

Answer £ ....559.50....... *(5 marks)*

D 2 Here are six numbers written in order:

2   5   7   8   9   a   b

Work out the values of *a* and *b* so that:
- the mean of the six numbers is 11
- the range of the numbers is three times larger than the median.

Median = 8 $8 \times 3 = 24$ $24 + 2 = 26$ $b = 28$
$11 \times 6 = 66 - 49 = 21 = a$
$2 + 5 + 7 + 9 + 26 = 49$

Answer *a* = ...17....   *b* = ...28.... *(4 marks)*

D 3 Paulo is buying a new car and selling his old car.
Here are three offers for the same model of new car.

Offer 1	Offer 2	Offer 3
New car £12 550	New car £13 068	New car £312.50 per
If old car given to	If old car given to	month for 3 years
garage, 30% off.	garage, $\frac{1}{3}$ off.	If old car given to garage, get £2500 cashback.

Which is the cheaper offer?
You **must** show your working.

O1 $12,550 \times 0.7 = 8785$
O2 $13068 \div 3 = 4356 \times 2 = 8712$ 8712
O2 $312.5 \times 36 = 11250 - 2500 = 8750$

Answer Offer ....2.... is cheapest *(5 marks)*

C 4 Two bags of sweets each only contain blackjacks and fruities.
There are fewer than 50 sweets in bag A.
5% of the sweets are blackjacks.
Write down **all** possible values for the number of sweets in bag A.

$5\% = 1$ $95\% = 19$ Bag A = 20, 40
$5\% = 2$ $95\% = 38$ Answer ....20, 40.... *(2 marks)*

Problem-solving practice 2

B 5 In 2011 around 1 200 000 000 000 litres of water were lost in the UK through leakages.

(a) Write this number in standard form.

Answer ....1.2×10^{12} litres........ *(1 mark)*

(b) This amount is 7% higher than the water loss in 2001.
Calculate the water loss in 2001.
Give your answer in standard form correct to 3 significant figures.

$1\% = 1.2 \times 10^{10}$ $100\% - 7\% = 93\%$
$7\% = 8.4 \times 10^{10}$ $93\% = 1.12 \times 10^{12}$

Answer ....1.12×10^{12} litres........ *(3 marks)*

B 6 The box plots show information about the mass, in grams, of eggs collected at two different farms.

$\cancel{B}A = 25\% \leqslant 59g \quad 75\% \geqslant 69g$
Median $= 63g$. $IQR = 10g \quad R = 27$

$\cancel{A}B = 25\% \leqslant 6\cancel{k} \quad 75\% \geqslant 68g$
Median $= 63g \quad IQR = 7g \quad R = 20$

An egg must weigh more than 63 g to be classified as Large.
One egg is taken at random from each farm.
Work out an estimate of the probability that both eggs are classified as Large.

$\%$ of eggs $\leqslant 63g \quad A = LQ \div 2 = 12.5\% +$ Median $= 62.5\%$
$(63 - \text{Lower Extreme}) \div \text{Range} \quad B = \text{Median} = 50\%$ $50\% \times 62.5\% = 31.25\%$
0.3125

Answer ....0.3125........ *(3 marks)*

(Farm A: / Farm B: box plots on grid, Mass (g) axis 45 to 80)

A 7 Andrew has these coins:

20   20   10   10 $\cancel{2}$ 1 $63p$

Samir has these coins:

50   10 $\cancel{1}$ $61p$

Andrew takes one of his coins at random and gives it to Samir.
Samir adds it to his coins.
Then Samir takes one of his coins at random and gives it to Andrew.
What is the probability that Andrew and Samir now have the same amount of money as each other?
You **must** show your working.

$\cancel{12} \quad \dfrac{1}{6} \times \dfrac{1}{4} = \dfrac{1}{24}$

Answer ....$\dfrac{1}{24}$........ *(4 marks)*

Factors and primes

1 Write 112 as the product of its prime factors.
Give your answer in index form.

Guided

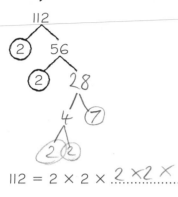

> Try to divide by the prime numbers starting with the smallest. The prime numbers are 2, 3, 5, 7, 11, 13, …

112 = 2 × 2 × $2 \times 2 \times 7$ ....

> Make sure that you give your final answer in index form.

Answer $2^4 \times 7$ .... *(3 marks)*

2 (a) Write 160 as the product of its prime factors.
Give your answer in index form.

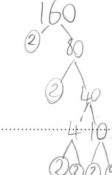

$2 \times 2 \times 2 \times 2 \times 2 \times 5$ ....

Answer $2^5 \times 5$ .... *(3 marks)*

(b) Work out the highest common factor (HCF) of 160 and 24.

> Circle all the prime numbers which are common to both products of prime factors. Multiply the circled numbers together to give the HCF.

$160 = 2 \times 2 \times 2 \times 2 \times 5$
$24 = 2 \times 2 \times 3 \times 2$

$2 \times 2 \times 2 = 8$

Answer ...... 8 ...... *(2 marks)*

(c) Work out the lowest common multiple (LCM) of 160 and 24.

> Multiply the HCF by all of the factors not circled in part **(b)**.

$2 \times 2 \times 3 \times 5 = 60 \times 8 = 480$

Answer ...... 480 ...... *(2 marks)*

3 p is a number.
8 is the highest common factor of 16 and p.
Work out **two** different possible values for p.

> Use the multiples of 8.

$8 = 2 \times 2 \times 2$

$16 = 2 \times 2 \times 2$ $p = 2 \times 2 \times 2 \times 3$ or $2 \times 2 \times 2 \times 5$

Answer ...... 24 ...... and ...... 40 ...... *(2 marks)*

Fractions and decimals

D **1** Write 0.77, $\frac{17}{20}$ and 79% in order with the smallest first.

> Convert the fraction and percentage into decimals.

Guided

$$\frac{17}{20} \overset{\times 5}{=} \frac{85}{100} = 0.85 \quad 100 \div 79\% = 0.79$$

0.77   0.85

Answer ...... 0.77 0.79 0.85 ...... *(2 marks)*

D **2** Write 0.11, $\frac{3}{40}$ and 9% in order with the smallest first.

0.11   9% ÷ 100 = 0.09 $\frac{3}{40}$ = 3 ÷ 40 = 0.075

0.075
40⟌300
   200

Answer ..$\frac{3}{40}$.. 0.09 0.11 ...... *(2 marks)*

D **3** Work out $\frac{5}{6} - \frac{4}{9}$

> Make sure that the denominators are the same before subtracting.

Guided

$$\frac{5}{6} = \frac{15}{18} \qquad \frac{4}{9} = \frac{8}{18} \qquad \frac{5}{6} - \frac{4}{9} = \frac{15}{18} - \frac{8}{18} = \frac{7}{18}$$

Answer ......$\frac{7}{18}$...... *(2 marks)*

D **4** Work out $\frac{4}{7} + \frac{2}{3}$

Give your answer as a mixed number.

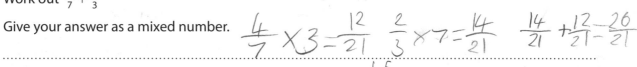

$$\frac{4}{7} \times 3 = \frac{12}{21} \qquad \frac{2}{3} \times 7 = \frac{14}{21} \qquad \frac{14}{21} + \frac{12}{21} = \frac{26}{21}$$

Answer ......$1\frac{5}{21}$...... *(2 marks)*

C **5** Work out $5\frac{1}{2} + 2\frac{1}{8}$

Give your answer as a mixed number.

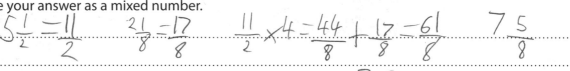

$$5\frac{1}{2} = \frac{11}{2} \qquad 2\frac{1}{8} = \frac{17}{8} \qquad \frac{11}{2} \times 4 = \frac{44}{8} + \frac{17}{8} = \frac{61}{8} \qquad 7\frac{5}{8}$$

Answer ......$7\frac{5}{8}$...... *(3 marks)*

C **6** Work out $1\frac{1}{6} + 1\frac{5}{8}$

Give your answer as a mixed number in its simplest form.

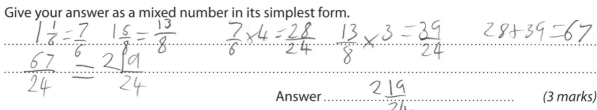

$$1\frac{1}{6} = \frac{7}{6} \qquad 1\frac{5}{8} = \frac{13}{8} \qquad \frac{7}{6} \times 4 = \frac{28}{24} \qquad \frac{13}{8} \times 3 = \frac{39}{24} \qquad 28 + 39 = 67$$

$$\frac{67}{24} = 2\frac{19}{24}$$

Answer ......$2\frac{19}{24}$...... *(3 marks)*

C **7** Work out $4\frac{1}{6} - 1\frac{4}{5}$

Give your answer as a mixed number in its simplest form.

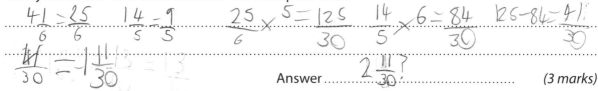

$$\frac{41}{6} = \frac{25}{6} \qquad 1\frac{4}{5} = \frac{9}{5} \qquad \frac{25}{6} \times 5 = \frac{125}{30} \qquad \frac{14}{5} \times 6 = \frac{84}{30} \qquad 125 - 84 = 41$$

$$\frac{41}{30} = 1\frac{11}{30} = 13$$

Answer ......$2\frac{11}{30}$?...... *(3 marks)*

Decimals and estimation

D 1 Given that $23.6 \times 41 = 967.6$

(a) Work out $967.6 \div 4.1$

> Use an inverse operation, but be careful as the denominator is 10 times smaller, so the answer will be 10 times larger.

Answer .... 236 .... *(1 mark)*

(b) Work out 0.236×410

> 23.6 has been divided by 100, and 41 has been multiplied by 10. You need to divide the answer by 100 and then multiply by 10.

Answer .... 96.76 .... *(1 mark)*

D 2 You are given that $32.5 \times 24 = 780$

(a) Work out 32.5×2.4

Answer .... 78 .... *(1 mark)*

(b) Work out $\frac{780}{240}$ $780 \div 240 = 3.25$

Answer .... 3.25 .... *(1 mark)*

(c) Work out 32.5×25

$$\begin{array}{r} 780 \\ 32.5 \\ \hline 812.5 \end{array}$$

> Notice that the 24 in the original calculation is now 25. You have one extra lot of 32.5, so add 32.5 to the original answer.

Answer .... 812.5 .... *(1 mark)*

D 3 Use approximations to estimate the value of $\frac{\sqrt{95.9}}{2.21}$

> Round both numbers to one significant figure.

$$\frac{\sqrt{100}}{2} = \frac{10}{2} = 5$$

Answer .... 5 .... *(2 marks)*

C 4 Use approximations to estimate the value of $\frac{3.1 \times 205}{0.521}$

 EXAM ALERT

> Students have struggled with exam questions similar to this – **be prepared!**

> 0.521 to 1 significant figure is 0.5, **not** 1.

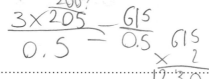

$$\frac{3 \times 205}{0.5} = \frac{615}{0.5} \quad \frac{615}{\times 2} \\ \hline 1230$$

Answer .... 1230/1200 .... *(3 marks)*

C 5 Use approximations to estimate the value of $\frac{9.81^2}{0.411}$

$$\frac{10^2}{0.4} = \frac{100}{0.4} = 100 \times 2.5 = 250$$

Answer .... 250 .... *(3 marks)*

C 6 Use approximations to estimate the value of $\frac{809.5}{10.2 \times 1.9^2}$

$$\frac{800}{10 \times 2^2} = \frac{800}{40} = 80 \div 4 = 20$$

Answer .... 20 .... *(3 marks)*

Recurring decimals

C **1** Show that $\frac{7}{20}$ can be written as a terminating decimal.

> $5 \times 20 = 100$, so multiply top and bottom by 5, then convert to a decimal.

$\frac{7}{20} \times 5 = \frac{35}{100} = 0.35$

(1 mark)

C **2** Show that $\frac{7}{80}$ can be written as a terminating decimal.

> You **could** divide 7 by 80, but it's probably easier to write the denominator as a product of its prime factors and use that in your explanation.

80
8 40
     4 10
Only uses 2 and 5.

(2 marks)

C **3** Show that $\frac{12}{99}$ is a recurring decimal.

> You **could** divide 12 by 99, but it's probably easier to write the denominator as a product of its prime factors and use that in your explanation.

99
3 33
    3 11
Uses primes other than 2 and 5.

(2 marks)

C **4** Show that $\frac{7}{12}$ can **not** be written as a terminating decimal.

$12 = 2 \times 2 \times 3$ Uses primes other than 2 and/or 5

(2 marks)

C **5** Write $\frac{4}{7}$ as a recurring decimal.

> Use long division and make sure you show enough decimal places to show the recurring decimal.

$0.\dot{5}7142\dot{8}$

Answer $0.\dot{5}7142\dot{8}$

(2 marks)

C **6** **(a)** Show that $\frac{2}{3}$ is equivalent to 0.666…

> Use long division to divide 2 by 3.

$3\overline{)20}$ .666
$0.\dot{6}$

(1 mark)

(b) Hence, or otherwise, write 0.7666… as a fraction.

> Work out the difference between 0.666… and 0.7666…. Convert that decimal to a fraction, then add it on to the $\frac{2}{3}$.

$0.76 - 0.66 = 0.1$
$0.1 = \frac{10}{100} - \frac{1}{10} \times \frac{3-3}{30} \frac{20}{30} - \frac{23}{30}$

Answer $\frac{23}{30}$

(3 marks)

29

Percentage change 2

1 Joe sees the same model of camera for sale in two different shops.

Shop A	**Shop B**
£150 (+ 20% VAT)	Sale! £190 now 5% off!
180	*180.5*

> **Guided**

For Shop A, find 10% and double it. For Shop B find 10% and halve it.

In which shop is the camera cheaper?
You **must** show your working.

Shop A: 10% of £150 = £15

20% of £150 = £30......

5% of 190 = 9.5   Total price = £150 + £ ...30..... = £ ...180...

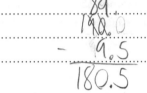

180.5

Answer ......£180.5.............   *(4 marks)*

2 Two banks calculate the yearly interest they pay customers.

Bank A	**Bank B**
3% of the total that you invest	1% of the first £300 that you invest
For example: Invest £1000	5% of amounts over £300 that you invest
Interest = 3% of £1000	For example: Invest £1000
	Interest = 1% of £300 + 5% of £700

Billy has £500 to invest for one year.
Work out which bank will pay him more interest.
State how much **extra** interest he will earn.

A   500×1.03   B 1% of 300 = 3   B = 513
      1% of 500 = 5   5% of 200 = 10
A = 515   B 10
              + 3
              13   Answer ....A. £2 extra.........   *(4 marks)*

3 Last year, 7 students went to a theme park. The total cost of the tickets was £140.
This year, 10 students are going.
The cost of each ticket has increased by 15%.
They have a total of £240.
Is this enough to buy 10 tickets?
You **must** show your working.

> **Guided**

Last year, each ticket cost = £140 ÷ 7 = £ ..20...

Increase in cost of each ticket = 15% of ..20.. = 10% of ..20.. + 5% of ..20.. = £ ..23..

23 ×10 = 230

Answer ....Yes, cost is £230.............   *(5 marks)*

Ratio problems

D

Guided

1 A car costs £7000.
Alice and Juan share the cost in the ratio 2 : 3.
Alice has saved £120 each month for the last two years.
Can she pay for all of her share using her savings?
You **must** show your working.

> Work out Alice's share of the £7000 and then work out how much she has saved.

Number of parts = 2 + 3 = 5

Each part = £7000 ÷ ....5 = 1400............

Alice's share = 2 × .....1400.... = ...2800......

$120 \times 24 = 2880$
£2400
481
2880

Answer ......Yes, she has £2880, so she has £80 over...... *(4 marks)*

D

2 Here is a list of ingredients.
Greg is making a pudding for 14 people using these ingredients.
Work out the number of grams of pudding rice he needs.

Serves 4 people	
Butter	20 g
Pudding rice	60 g
Sugar	30 g
Sultanas	40 g
Milk	700 ml

> Work out what mass of rice is needed for one person and multiply by 14.

60 ÷ 4 = 15

15
× 14
60
150
210

Answer ....210............... g   *(3 marks)*

C

Guided

3 Concrete is made by mixing cement, sand and gravel in the ratio 1 : 2 : 5.
A builder mixes 360 kg of concrete.
How much gravel does the builder need?

Total number of parts = 1 + 2 + 5 = 8   45 ÷ 5 = 9

Each part = 360 kg ÷ 8 = ...45kg....

45
× 5
225

Gravel = ...225........... kg   *(3 marks)*

B

EXAM ALERT

4 A paint manufacturer mixes blue and white in two different ratios to make different shades.

> Students have struggled with exam questions similar to this – **be prepared!**

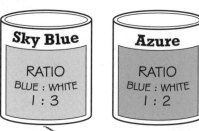

Sky Blue
RATIO
BLUE : WHITE
1 : 3

Azure
RATIO
BLUE : WHITE
1 : 2

Jamie has 600 ml of Sky blue. How much extra blue paint should he add to turn it into Azure?

SB   1:3 × 2 = 2:6   600 ÷ 8   8)600   75
A   1:2 × 3 = 3:6   40

Answer ....75............... ml   *(4 marks)*

Indices 1

1 Work out

    (a) $(-1)^1$ Answer -1 *(1 mark)*

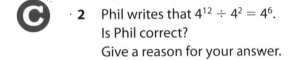

    (b) $(-1)^2$ Answer $-1 \times -1 =$ ....   *(1 mark)*

    (c) $(-1)^3$ Answer $-1 \times -1 \times -1 =$ ....   *(1 mark)*

    (d) $(-1)^4$ Answer $-1 \times -1 \times -1 \times -1 =$ ....   *(1 mark)*

    (e) Using your answers to a, b, c and d above, or otherwise, calculate $(-1)^{11}$.

         > Look at the pattern made by your answers.

    ....

         Answer ....   *(1 mark)*

2 Phil writes that $4^{12} \div 4^2 = 4^6$.
Is Phil correct?
Give a reason for your answer.

     > You can give your reason by calculating the correct value of $4^{12} \div 4^2$.

....

....   *(2 marks)*

3 Write each expression as a single power.

     > $a^m \times a^n = a^{m+n}$
     > $a^m \div a^n = a^{m-n}$

    (a) $2^3 \times 2^3$

         Answer ....   *(1 mark)*

    (b) $\dfrac{5^7}{5^5}$

         Answer ....   *(1 mark)*

    (c) $7^8 \div 7^2$

         Answer ....   *(1 mark)*

4 **(a)** Write $2^3 \times 2^6 \times 2^{-5}$ as a single power of 2.

    ....

         Answer ....   *(2 marks)*

    (b) Write $\dfrac{5^2 \times 5^5}{5^4}$ as a single power of 5.

    ....

         Answer ....   *(2 marks)*

5 Write each expression as a single power.

    (a) $3^{-5} \times 3^3$

         Answer ....   *(1 mark)*

    (b) $\dfrac{9^2}{9^6}$

         Answer ....   *(1 mark)*

    (c) $(2^4)^3$

         Answer ....   *(1 mark)*

Indices 2

B **1** Write 3^{-3} as a fraction.

 Guided

$\dfrac{1}{3^3}$

> Do not leave indices in your fraction. Write down the value of 3^3.

Answer .... *(1 mark)*

B **2** Write down the value of $\left(\dfrac{2}{3}\right)^{-1}$.

> Raising to the power of -1 is the same as taking the reciprocal. Turn the fraction upside down.

Answer .... *(1 mark)*

B **3** Here are two numbers: 2^{-3} 1.5×10^{-1}
Which is the larger?
You **must** show your working.

....

Answer .... *(3 marks)*

A **4** Work out the value of $\left(\dfrac{2}{5}\right)^{-2}$

> Turn the fraction upside down, then square the numerator and the denominator.

....

Answer .... *(2 marks)*

A **5** Work out the value of each of the following:

(a) $27^{\frac{2}{3}}$

> **(a)** Work out the cube root first, then square the result.

Answer .... *(2 marks)*

(b) $100^{\frac{3}{2}}$

Answer .... *(2 marks)*

(c) $64^{\frac{1}{3}}$

Answer .... *(2 marks)*

(d) $16^{-\frac{1}{2}}$

> **(d)** The index is negative, so start by taking the reciprocal.

Answer .... *(2 marks)*

A **6** $\dfrac{25^{0.5} \times 5^4}{25} = k$ where k is an integer.

> Write all the numbers as powers of 5 first.

Work out k.

....

....

$k =$ .... *(2 marks)*

A* **7** $x = (0.5)^{-4}$ $y = 27^{\frac{1}{3}}$
Work out the value of xy.

....

....

$xy =$ .... *(4 marks)*

Surds

Guided

1 Simplify the following expressions.

(a) $\sqrt{7} \times \sqrt{3} = \sqrt{7 \times 3} = \sqrt{....?.!....}$

> **(a)** Use the rule $\sqrt{ab} = \sqrt{a} \times \sqrt{b}$.

Answer .... *(1 mark)*

(b) $\sqrt{21} \div \sqrt{3}$

> **(b)** Use the rule $\sqrt{\dfrac{a}{b}} = \dfrac{\sqrt{a}}{\sqrt{b}}$.

Answer .... *(1 mark)*

(c) $\dfrac{\sqrt{18}}{\sqrt{6}}$

Answer .... *(1 mark)*

> **(d), (e)** Simplify each term and then simplify the expression.

(d) $\sqrt{18} + \sqrt{8} = 3\sqrt{2} +$ ....

Answer .... *(2 marks)*

(e) $\sqrt{75} - \sqrt{12}$

Answer .... *(2 marks)*

2 Write $\dfrac{8}{\sqrt{2}} + \sqrt{18}$ in the form $a\sqrt{2}$, where a is an integer.

> Start by rationalising the denominator of $\dfrac{8}{\sqrt{2}}$ and simplifying the $\sqrt{18}$.

....

....

Answer .... *(4 marks)*

3 Work out values of a and b such that $\left(3 - \sqrt{2}\right)^2 = a + b\sqrt{2}$.

....

Answer .... *(3 marks)*

4 Here is a formula: $r = \sqrt{p^2 - q^2}$

Find the value of r when $p = 5\sqrt{3}$ and $q = \sqrt{15}$.

Write your answer in the form $a\sqrt{b}$ where a and b are integers greater than 1.

....

....

$r =$ .... *(3 marks)*

5 Work out the value of x if $\dfrac{x\sqrt{2}}{3 - \sqrt{3}} = 3 + \sqrt{3}$

> Start by multiplying both sides by $3 - \sqrt{3}$.

Give your answer in the form of $a\sqrt{b}$ where a and b are integers.

....

....

....

$x =$ .... *(4 marks)*

Algebraic expressions

 1 Simplify each of the following expressions:

 (a) $y^4 \times y^5$ $= y^{4+5} = y$.......

> Use the rule $a^m \times a^n = a^{m+n}$ *(1 mark)*

(b) $x \times x^3$

Answer .... *(1 mark)*

(c) $\dfrac{p^8}{p^2}$ $= p^{8-2} = p$.......

> Use the rule $\dfrac{a}{a^n} = a^{m+n}$ *(1 mark)*

(d) $t^{15} \div t^5$

Answer .... *(1 mark)*

(e) $(m^2)^5$

> Use the rule $(a^m)^n = a^{mn}$

Answer .... *(1 mark)*

 2 Simplify each of the following expressions:

(a) $y^{-4} \times y^{-5}$

Answer .... *(1 mark)*

(b) $x \times x^{-3}$

Answer .... *(1 mark)*

(c) $\dfrac{p^{-8}}{p^2}$

Answer .... *(1 mark)*

(d) $t^{15} \div t^{-5}$

Answer .... *(1 mark)*

 3 Simplify fully each of the following expressions:

 (a) $\dfrac{25a^8b^6}{5a^3b^4}$

Answer .... *(2 marks)*

(b) $6cd^2 \times 3c^3d^3$

Answer .... *(2 marks)*

(c) $(2pq^4)^3$ $= 2^3p^3(q^4)^3 =$ ....

Answer .... *(2 marks)*

(d) $(3d^3e)^3$

Answer .... *(2 marks)*

Ⓐ 4 Simplify the expression $(64n^6)^{\frac{1}{3}}$

....

Answer .... *(2 marks)*

Ⓐ 5 Simplify fully $\left(\dfrac{16x^6y^4}{2x^2y}\right)^2$

> Simplify the brackets first, then square each term of the simplified expression.

....

....

Answer .... *(4 marks)*

Expanding brackets

D **1** Multiply out each of the following brackets:

(a) $4(y + 5)$

> First work out $4 \times y$, then 4×5.

Guided

$4(y + 5) = 4 \times y + 4 \times 5 = $ .... *(2 marks)*

(b) $3(y^2 - y + 2)$

Answer .... *(2 marks)*

(c) $5(a + 2b - 3c)$

> Your expression should have three terms.

Answer .... *(2 marks)*

D **2** Expand $d(d - 5)$

> 'Expand' means the same as 'multiply out'.

Answer .... *(2 marks)*

D **3** Show that $2(3x + 2) + 5x = 11x + 4$

> Expand and simplify the left-hand side.

....

.... *(2 marks)*

D **4** Show that $3(x + 2) + 3x = 2(3x + 3)$

> Expand and simplify both sides of the equation.

....

....

.... *(2 marks)*

C **5** Multiply out each of the following brackets:

(a) $5y(y - 6)$

Answer .... *(2 marks)*

(b) $2x^2 (2x - 5)$

Answer .... *(2 marks)*

(c) $5y^2(y + x)$

Answer .... *(2 marks)*

C **6** Expand and simplify $(y + 7)(y - 2)$

> You can use FOIL or a grid. Remember to simplify.

....

....

Answer .... *(2 marks)*

B **7** Expand and simplify $(2n - 5)^2$

> Students have struggled with exam questions similar to this – **be prepared!**
>

EXAM ALERT

....

....

Answer .... *(2 marks)*

Factorising

 1 Factorise fully $4t - 20$ Answer $4(\ldots\ldots\ldots - \ldots\ldots\ldots)$ *(1 mark)*

> The highest common factor of $4t$ and 20 is 4.

Guided

 2 Factorise fully $9 + 21p$ Answer $\ldots\ldots\ldots\ldots\ldots\ldots\ldots\ldots$ *(1 mark)*

> The highest common factor of 9 and $21p$ is 3.

C **3** Factorise the following expressions fully:

> The highest common factor of $2a^2$ and $8a$ is $2a$.

Guided

  (a) $2a^2 - 8a$

Answer $2a(\ldots\ldots\ldots\ldots\ldots\ldots\ldots\ldots$ *(1 mark)*

  (b) $n^2 + 11n + 30$

> You need to write this expression with **two** brackets. Think of two numbers which multiply to give 30 and add up to 11.

$\ldots\ldots\ldots\ldots\ldots\ldots\ldots\ldots\ldots\ldots\ldots\ldots\ldots\ldots\ldots\ldots\ldots\ldots$

Answer $\ldots\ldots\ldots\ldots\ldots\ldots\ldots\ldots\ldots$ *(2 marks)*

C **4** $2(x + 8) + 3(x - 2)$ simplifies to $a(x + b)$ Work out the values of a and b.

> Expand the brackets, simplify and then factorise. Remember to give the values of a and b.

$\ldots\ldots\ldots\ldots\ldots\ldots\ldots\ldots\ldots\ldots\ldots\ldots\ldots\ldots\ldots\ldots\ldots\ldots$

Answer $a = \ldots\ldots\ldots\ldots\ldots$ $b = \ldots\ldots\ldots\ldots\ldots$ *(3 marks)*

B **5** Factorise the following expressions fully:

  (a) $10d^2 - 15de$

Guided

Answer $5d(\ldots\ldots\ldots\ldots\ldots\ldots\ldots\ldots$ *(2 marks)*

  (b) $10x^2 + 8xy$

Answer $\ldots\ldots\ldots\ldots\ldots\ldots\ldots\ldots\ldots$ *(2 marks)*

B **6** Factorise $x^2 - 81$

> Use the difference of two squares rule.

$\ldots\ldots\ldots\ldots\ldots\ldots\ldots\ldots\ldots\ldots\ldots\ldots\ldots\ldots\ldots\ldots\ldots\ldots$

Answer $\ldots\ldots\ldots\ldots\ldots\ldots\ldots\ldots\ldots$ *(2 marks)*

A **7** Factorise $16x^2 - 25$

> Use the difference of two squares rule.

$\ldots\ldots\ldots\ldots\ldots\ldots\ldots\ldots\ldots\ldots\ldots\ldots\ldots\ldots\ldots\ldots\ldots\ldots$

Answer $\ldots\ldots\ldots\ldots\ldots\ldots\ldots\ldots\ldots$ *(2 marks)*

A **8** Factorise $7x^2 - 13x - 2$

$\ldots\ldots\ldots\ldots\ldots\ldots\ldots\ldots\ldots\ldots\ldots\ldots\ldots\ldots\ldots\ldots\ldots\ldots$

$\ldots\ldots\ldots\ldots\ldots\ldots\ldots\ldots\ldots\ldots\ldots\ldots\ldots\ldots\ldots\ldots\ldots\ldots$

Answer $\ldots\ldots\ldots\ldots\ldots\ldots\ldots\ldots\ldots$ *(2 marks)*

Algebraic fractions

 1 Simplify fully $\dfrac{9}{7a} + \dfrac{3}{14a}$

> Make both denominators $14a$. Add the two fractions, then cancel down if possible.

Answer .... *(3 marks)*

 2 Simplify fully $\dfrac{1}{x+3} + \dfrac{5}{x+4}$

> Convert the two fractions to ones with a common denominator of $(x+3)(x+4)$ by multiplication. Remember to simplify the numerator after multiplying.

Guided

$$= \frac{1(x+4)}{(x+3)(x+4)} + \frac{5(x+3)}{(x+3)(x+4)} = \frac{\dots\dots\dots\dots\dots}{(x+3)(x+4)}$$

Answer .... *(3 marks)*

 3 Simplify fully $\dfrac{3}{4x+1} - \dfrac{4}{x+2}$

Answer .... *(3 marks)*

 4 Simplify fully $\dfrac{x}{3} \div \dfrac{5}{x-2}$

Answer .... *(2 marks)*

 5 Simplify fully $\dfrac{x+5}{5} \div \dfrac{2x+10}{3}$

Answer .... *(3 marks)*

 6 Simplify fully $\dfrac{x^2 - 2x - 8}{x^2 - 4}$

EXAM ALERT

> Students have struggled with exam questions similar to this – **be prepared!**

Answer .... *(3 marks)*

 7 Simplify fully $\dfrac{9x^2 - 64}{3x^2 + 2x - 16}$

Answer .... *(3 marks)*

Straight-line graphs

 1 On the grid, draw the graph of:

 (a) $y = 2x - 1$ *(3 marks)*

x	-4	-2	O	2	4
y	-9				

(b) $y = x$

.... *(2 marks)*

(c) $y = \frac{1}{2}x + 6$

.... *(3 marks)*

 (d) $2x + y = 2$

> Students have struggled with exam questions similar to this – **be prepared!**

.... *(3 marks)*

(e) $5x + 2y = 20$

.... *(3 marks)*

> A table often helps. Choose some values for x. Substitute each value of x into the equation to find the value of y.

B **2** Look at the grid below.

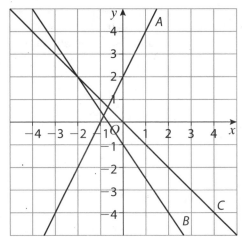

(a) Work out the equation of line *A* in the form $y = mx + c$

Answer.. *(2 marks)*

(b) Work out the equation of line *B*.

Answer.. *(2 marks)*

(c) Work out the equation of line *C*.

Answer.. *(2 marks)*

(d) Line *D* is parallel to line *A* and passes through the point $(0, -3)$.
Write down the equation of the line in the form $y = mx + c$

Answer.. *(2 marks)*

> Draw a triangle on line *A* to work out the gradient *m*. *c* is the point where the line crosses the *y*-axis.

> Parallel means that their gradients will be the same.

Gradients and midpoints

1 *P* is (0, 4), *Q* is (7, 7).
Work out the coordinates of the midpoint of the line segment *PQ*.

> The coordinates (x, y) of the midpoint is the mean of the two x-coordinates and the mean of the two y-coordinates.

....

Answer (............., ...)   *(2 marks)*

2 Two straight lines are shown.

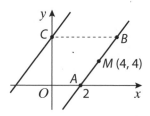

> You need to show that the lines are parallel. Work out the gradient for both lines – if they are the same, they are parallel.

Prove that the lines never meet.

Gradient of A = 2 ÷ 4 = ...

Gradient of B = ...

....   *(3 marks)*

3 On the grid, *A* is the point (2, 0).
The midpoint, *M*, of *AB* is (4, 4).
B and *C* are on the same horizontal line.

> Work out the gradient of the line through *A* and *B*. The line through *C* will have the same gradient.

Work out the equation of the line through *C* that is parallel to *AB*.

....

....

Answer ....   *(3 marks)*

4 The graph shows two lines *A* and *B*.
The equation of line *A* is $3x - 4y + 12 = 0$
Line *A* crosses the axes at *P* and *Q*.
Q is the mid-point of *PR*.
Line *B* passes through the point *S* (2, 1).
Line *B* crosses the *y*-axis at *T*.
S is the mid-point of *RT*.
Work out the equation of line *B*.

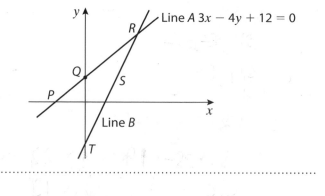

....

....

Answer ....   *(3 marks)*

Real-life graphs

1   Plan A and Plan B are two monthly mobile phone plans with the same free phone.
Here are the details of Plan A. Monthly charge £36.
300 minutes of calls free. Each extra minute 30p.
Plan B is a different mobile phone tariff.
These graphs show the cost, in £, for each of the plans.

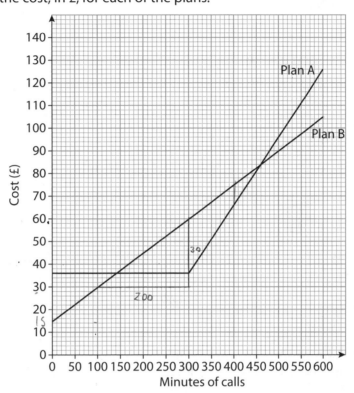

(a) What is the monthly charge for plan B?

Answer .... (1 mark)

(b) How many free minutes are included in Plan B?

Answer .... (1 mark)

(c) Bertha chooses Plan B.
How much does she pay for each
minute of calls?

> Cost per minute is the gradient of the line. You can
> draw a triangle if it helps.

....

Answer .... (2 marks)

(d) Andy usually makes about 400 minutes of calls a month. Which plan should he choose?

Answer .... (1 mark)

(e) Give a reason for his choice.

....

Answer .... (1 mark)

(f) After how many minutes of calls are the two plans the same cost?

....

Answer .... (1 mark)

Formulae

1 $w = -2$, $x = -6$ and $y = 4$

Guided

Work out the value of $\dfrac{xy}{wx - y}$

> Use brackets when you substitute negative numbers.

$$\frac{(-6) \times 4}{-2 \times (-6) - 4} = \frac{\cdots\cdots\cdots}{\cdots\cdots\cdots} = \cdots\cdots\cdots$$

Answer ....   *(3 marks)*

2 $a = 10$, $b = 3$ and $c = -6$

Guided

Work out the value of $\dfrac{ab - c}{c + 8}$

$$\frac{10 \times 3 - (-6)}{(-6) + 8} = \frac{\cdots\cdots\cdots}{\cdots\cdots\cdots} = \cdots\cdots\cdots$$

> $10 \times 3 - (-6) = 10 \times 3 + 6$

Answer ....   *(3 marks)*

3 $p = 3$, $q = -5$ and $r = 4$

Work out the value of w when $w = \dfrac{6(p - q)}{r^2}$

....

....

Answer ....   *(3 marks)*

4   The graph shows the cost, C (£), of hiring a lorry from Red Lorry Hire for a number of days, d.

> $y = mx + c$ should be used to help, where m is the gradient and c is the intercept.

(a) Circle the correct formula for the cost, C, of hiring a lorry from Red Lorry Hire.

   $C = 50d$ $C = 120d$

   $C = 70d + 50$ ✓ $C = 50d + 70$

                        (1 mark)

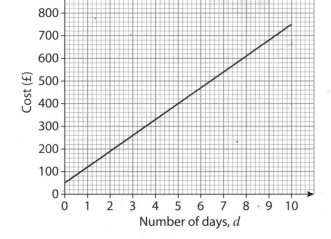

(b) Another firm, Yellow Lorry Hire, charge their customers using this formula
$C = 40d + 260$
Plot the graph of $C = 40d + 260$ on the graph paper above.

> Substitute $d = 0$ into the equation to work out one coordinate. Use $d = 10$ to work out another coordinate.

....

....

....   *(2 marks)*

(c) After how many days is it cheaper to use Yellow Lorry Hire than Red Lorry Hire?

                Answer .... days   *(2 marks)*

Linear equations 1

1 Solve each of the following:

(a) $5(x + 3) = 25$

$5x + 15 = 25$

> Next get rid of the + 15 by subtracting 15 from both sides.

....

$x =$ .... *(3 marks)*

(b) $7(x - 2) = 49$

....

....

$x =$ .... *(3 marks)*

(c) $4(2x - 3) = 28$

....

....

$x =$ .... *(3 marks)*

(d) $2x + 3 = x + 10$

> Collect like terms. Start by getting rid of the smallest x term by subtracting x from both sides.

....

$x =$ .... *(2 marks)*

(e) $5x - 20 = 8 - 2x$

> $- 2x$ is the smallest x term, so get rid of it first.

....

....

$x =$ .... *(3 marks)*

(f) $\frac{e}{5} = -2$

$e =$ .... *(1 mark)*

2 Solve each of the following:

(a) $5x + 8 = 4(x + 3)$

....

....

$x =$ .... *(3 marks)*

(b) $\frac{3x - 1}{2} = 7$

....

....

$x =$ .... *(3 marks)*

43

Linear equations 2

Solve each of the following equations.

1 $\dfrac{12 - x}{3} = 6 - x$

> **Guided**

$12 - x = 18 - 3x$

> Both sides have been multiplied by 3. Next collect like terms by adding $3x$ to both sides.

> Students have struggled with exam questions similar to this – **be prepared!**

EXAM ALERT

....

....

$x = $ ....   *(3 marks)*

2 $\dfrac{x - 2}{3} - \dfrac{2x}{7} = -1$

> **Guided**

$21\left(\dfrac{x - 2}{3}\right) - 21\left(\dfrac{2x}{7}\right) = -1 \times 21$

> 21 is the LCM of 3 and 7. Next, cancel the fractions and multiply out the brackets.

....

....

....

$x = $ ....   *(4 marks)*

3 $\dfrac{x + 1}{2} + \dfrac{x + 2}{5} = 3$

> 10 is the LCM of 2 and 5.

....

....

....

....

$x = $ ....   *(4 marks)*

4 $\dfrac{3x + 1}{5} - \dfrac{5x - 1}{7} = 0$

> Be careful with the negative signs.

....

....

....

....

$x = $ ....   *(4 marks)*

5 $\dfrac{x - 1}{2} + \dfrac{3x + 4}{3} = 4$

....

....

....

....

$x = $ ....   *(4 marks)*

Number machines

1 This function machine multiplies a number by 2 then adds 5.

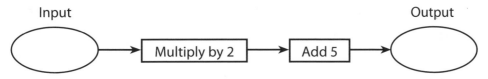

Input　　　　　　　　　　　　　　　　　　Output

Multiply by 2　→　Add 5

> The first part multiplies by 2, the second part adds 5 to the result.

(a) Work out the output when the input is 3.

Output = 3 × 2 + 5 = ... *(1 mark)*

(b) Work out the input when the output is 17.

Input = $\dfrac{17-5}{2}$ = ... *(1 mark)*

(c) Using x as the input, write the output as an expression.

Answer .... *(1 mark)*

EXAM ALERT

2 This function machine multiplies a number by 4 then adds 3.

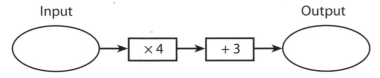

Input　　　　　　　　　　　　　Output

× 4　→　+ 3

> Students have struggled with exam questions similar to this –
> **be prepared!**

There is one value that gives an output that is twice the input. Work out this input.

> Use n to represent the input, then write an expression for the output. The output is twice the input. Use this fact to write an equation and solve it to find n.

....

....

....

....

Answer .... *(3 marks)*

3 Here is a number machine.

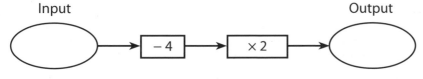

Input　　　　　　　　　　　　　　　　　Output

− 4　→　× 2

The output is three times the input.
Work out the input.

....

....

....

Answer .... *(3 marks)*

Inequalities

D

Guided

1   Write down the inequality that is represented by each of the following diagrams:

(a) ○————————→
    0   1   2   3

Answer $x >$ ....

(b) ●————————→
    6   7   8   9

Answer $x \geqslant$ ....

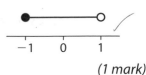
The open circle (○) means that the number under it is not included, so you must use < or >.
● means ⩽ or ⩾.

(c) ←————————●
    0   1   2   3

Answer 2

(d) ←————————○
    −4   −3   −2   −1

Answer ....   *(4 marks)*

(e) Circle the diagram that represents $-1 \leqslant n < 1$

●————————●   ○————————●   ○————————○   ●————————○
−1   0   1   −1   0   1   −1   0   1   −1   0   1

(1 mark)

C

2   n is an integer.
    List the values of n such that $-8 \leqslant 2n < 4$

    Answer ....   *(2 marks)*

C

3   Solve the inequality $2n - 1 > 9$

    Solve inequalities in the same way as you would solve equations, but here you must write > instead of =.

    ....

    Answer ....   *(2 marks)*

C

4   Solve the inequality $6x + 17 \leqslant 5$

    ....

    ....

    Answer ....   *(2 marks)*

C

5   Solve $8x > 3x + 20$

    ....

    ....

    Answer ....   *(2 marks)*

B

6   Solve $6x - 5 < 2x - 1$

    Students have struggled with exam questions similar to this – **be prepared!**

EXAM ALERT

    ....

    ....

    Answer ....   *(2 marks)*

B

7   Solve the inequality $3(n + 2) > 5n + 18$

    ....

    ....

    Answer ....   *(2 marks)*

Inequalities on graphs

1 Points in the shaded region satisfy three inequalities.
One of the inequalities is $y \leqslant 4$.
Which of these are the other **two** inequalities?

> Look at each inequality in turn. Work out where the line would be on the graph. If it matches one of the lines drawn, check whether the shading is on the correct side.

A $y > 2x - 3$ B $y < -2x + 3$

C $x + 3y \geqslant 9$ D $3y \leqslant 9 - x$

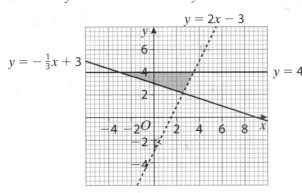

Answer .... and ....   *(2 marks)*

EXAM ALERT

2 On the grid below, shade the region satisfied by these three inequalities:

> Students have struggled with exam questions similar to this – **be prepared!**

$2y \geqslant x + 2$ $3x + y < 2$ $y \leqslant x + 3$

> Change each inequality into an = sign, and draw the graphs of each equation on the grid. It might help to rearrange each equation into the form $y = mx + c$.

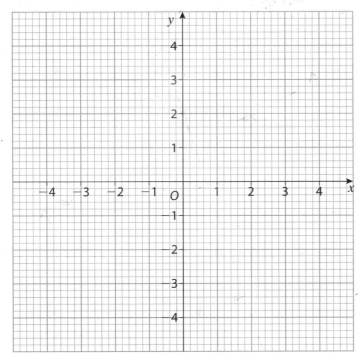

(4 marks)

Simultaneous equations 1

1 Solve the simultaneous equations:

$2a + b = 7$ 1
$5a - b = 14$ 2

> Number the equations to keep track of your working.
> Start by adding the equations to eliminate the b terms.

You **must** show your working. Do **not** use trial and improvement.

....

....

....

Answer ....   *(2 marks)*

2 Solve the simultaneous equations:

$5x + y = 2$ (1)
$x - 2y = -15$ (2)

You **must** show your working. Do **not** use trial and improvement.

$2 \times$ (1) $\longrightarrow$ $10x + 2y = 4$

$+$ (2) $\longrightarrow$ $x - 2y = -15$

....

....

Answer ....   *(3 marks)*

3 Solve the simultaneous equations:

$3x + 5y = 14$
$2x - 4y = -20$

> Multiply the first equation by 2, and the second
> equation by 3. Then subtract one equation from
> the other to eliminate the x terms.

You **must** show your working. Do **not** use trial and improvement.

....

....

....

Answer ....   *(4 marks)*

4 Three CDs and four DVDs cost £72. Five CDs
and two DVDs cost £64. What would three
CDs and five DVDs cost?
Do **not** use trial and improvement.
You **must** show your working.

> Your first job is to write out the information as a
> pair of simultaneous equations. The first one has
> been done for you. Make sure that you actually give
> the cost at the end for the three CDs and five DVDs.

$3c + 4d = 72$ ....

....

....

....

Answer ....   *(5 marks)*

Quadratic equations

 1   Solve the equation $x^2 - x - 6 = 0$

> **Guided**

$(x \dots\dots\dots\dots)(x \dots\dots\dots\dots) = 0$

> Which two numbers multiply to give -6 and add to give -1?

....

Answer.. *(3 marks)*

 2   Solve the equation $x^2 - 5x = 0$

> **Guided**

$x(\dots\dots\dots\dots\dots\dots) = 0$

> There is no number term, so one of the factors is just x.

....

Answer.. *(3 marks)*

 3   Solve the equation $x^2 - 36 = 0$

> There is no x term. This is a difference of two squares question.

....

....

Answer.. *(3 marks)*

 4   Solve the equation $4x^2 - 28x = 0$

....

....

Answer.. *(3 marks)*

 5   Solve the equation

EXAM ALERT

$3x(x + 6) = 4x + 5$

> Start by rearranging the equation into the form $ax^2 + bx + c = 0$.

> Students have struggled with exam questions similar to this – **be prepared!**

....

....

....

....

Answer.. *(3 marks)*

6   Two integers, a and b, are combined using the operation ✿ in the following way:

> **Guided**

$a ✿ b = a^2 + 3a - 3b - b^2$

> All that has happened is that x has been substituted for a and 2 has been substituted for b. Now simplify and solve as normal.

Find all solutions to the equation $x ✿ 2 = 0$

$x ✿ 2 = x^2 + 3x - 6 - 4 = 0$

....

....

Answer.. *(4 marks)*

Quadratics and fractions

1 (a) Show that $3 + \dfrac{4}{x+5} = \dfrac{6}{x}$ simplifies to $3x^2 + 13x - 30 = 0$

> The lowest common multiple (LCM) of the denominators is $(x+5)(x)$. Cancel down, expand the brackets, collect like terms, factorise and solve.

Guided

$$3(x+5)(x) + (x+5)(x)\frac{4}{x+5} = (x+5)(x)\frac{6}{x}$$

$3x^2 + 15x + 4x =$ ....

....

....

....

Answer.. *(3 marks)*

(b) Hence solve the equation:

$$3 + \frac{4}{x+5} = \frac{6}{x}$$

....

....

....

Answer.. *(3 marks)*

2 Solve $\dfrac{3}{2x-1} - \dfrac{4}{3x-1} = 1$

> The LCM of the denominators is $(2x-1)(3x-1)$.

....

....

....

....

Answer.. *(6 marks)*

3 Solve $\dfrac{8}{2x-1} - \dfrac{2}{x+1} = 2$

> Notice that all numerators are even – you could start by dividing by 2, or just use the same method as the previous two questions.

....

....

....

....

Answer.. *(6 marks)*

4 Solve $\dfrac{1+x}{3x+3} = \dfrac{1}{x+1}$

> The LCM of the denominators is $(3x+3)(x+1)$.

....

....

....

....

Answer.. *(6 marks)*

Simultaneous equations 2

Do **not** use trial and improvement to solve these equations. You **must** show your working.

1 Solve the simultaneous equations:

$y^2 = x + 3$ (1)

> **Guided**

$y = x - 3$ (2)

> Rearrange $y^2 = y + 3 + 3$ into the form $ay^2 + by + c = 0$, then factorise and solve. Remember to substitute your values of y into the easiest equation to work out the values of x.

$x = y + 3$ .... (3) ....

Substitute (3) into (1) ....

$y^2 = y + 3 + 3$ ....

....

....

....

Answer... *(6 marks)*

2 Solve the simultaneous equations:

$xy = 3$ (1)

> **Guided**

$y = x + 2$ (2)

> Do the substitution, expand the brackets, rearrange into the form $ax^2 + bx + c = 0$, factorise and solve.

Substitute (2) into (1) ....

....

....

....

....

....

....

Answer... *(6 marks)*

EXAM ALERT

3 Solve this pair of simultaneous equations:

$x^2 + y^2 = 25$

$y = 7 - x$

> Students have struggled with exam questions similar to this – **be prepared!**

....

....

....

....

....

....

Answer... *(6 marks)*

Rearranging formulae

C **1** Rearrange $y = 5x + 1$ to make x the subject.

$y - 1 = 5x$

> Use inverse operations – start by subtracting 1 from both sides. Then divide both sides by 5.

$x = $ .... *(2 marks)*

Guided

C **2** Make b the subject of $a = \frac{b}{3} + 2$

....

$b = $ .... *(2 marks)*

B **3** Make t the subject of the formula $2(t - p) = 3p + 4$

> Expand the brackets first.

....

....

$t = $ .... *(3 marks)*

B **4** Make r the subject of $p = q + \frac{r}{s}$

> Deal with the $+q$ before the $\div s$.

....

$r = $ .... *(2 marks)*

A **5** Rearrange $2(x + 2y) = 3(x - y)$ to make x the subject.

> Expand the brackets and then collect the x terms on one side of the formula.

....

....

$x = $ .... *(4 marks)*

A **6** Make a the subject of the formula

$3a - b = c(2 + a)$

> You will need to factorise to get a on its own.

....

....

....

....

$a = $ .... *(4 marks)*

A **7** Make x the subject of $y = \frac{x - 5}{x - 4}$

> Multiply both sides by the denominator.

....

....

....

....

$x = $ .... *(4 marks)*

Sequences 1

C **1** Here are the first three terms of an arithmetic sequence:

> Always look for the difference between the terms, then work out the zero term.

7   10   13   ………   ………

Which of the following is the nth term for this sequence?
Circle the correct answer.

$n + 3$ $3n + 7$ $3n + 4$ $7n + 3$ $n + 6$ *(1 mark)*

C **2** Here is an arithmetic sequence:

4   9   14   19   ………   ………

(a) Work out a formula for the nth term of the sequence.

....

nth term = ... *(2 marks)*

(b) Is 156 a term in this sequence?

> Look at the pattern made by the first few terms.

....

Answer .... *(2 marks)*

C **3** The nth term of a sequence is $4n - 3$.

> The term number, n, must always be an integer.

(a) Is 54 a term in this sequence?
You **must** show your working.

Guided

$4 \times 14 - 3 = 53$ $4 \times 15 - 3 = 57$

Answer .... *(2 marks)*

(b) How many terms of this sequence are **less than** 300?
You **must** show your working.

....

Answer .... *(2 marks)*

C **4** The nth term of a sequence is $n^2 - 3$.

(a) Is 67 a term in this sequence?

....

....

Answer .... *(2 marks)*

(b) What is the first term of this sequence that is greater than 100?

....

....

Answer .... *(2 marks)*

Sequences 2

 1 The nth term of a sequence is $2n - 4$.
Write down the first three terms of the sequence

> Substitute $n = 1$ into the nth term, then $n = 2$ and then $n = 3$.

Guided

Answer $-2,$ ...., ... *(2 marks)*

 2 The nth term of a sequence is $n^2 - 1$.
Write down the first three terms of the sequence.

Answer ...., ..., ... *(2 marks)*

 3 The nth term of a sequence is $\dfrac{3n - 2}{2}$

A term of the sequence is 17.
Work out the value of n for this term.

> Write an equation and solve it to find n.

Guided

$$\frac{3n - 2}{2} = 17$$

....

....

Answer .... *(2 marks)*

 4 The first three terms of a sequence are:

a b c ....

The rule for working out the next term in the sequence is:
Multiply the previous term by 3 and then subtract 4.
Show that $c = 9a - 16$

....

....

....

....

(4 marks)

 5 The rule for finding the next term in a sequence is to multiply the previous term by 2 then add on a where a is an integer.
The second term is 11 and the fourth term is 59.

.... 11 59

Work out the 1st term of the sequence:

The third term is $2 \times 11 + a$

The fourth term is $2 \times (2 \times 11 + a) + a = 59$

....

....

Answer .... *(5 marks)*

Algebraic proof

C **1** Harry says that when a and b are prime numbers, the value of $(3a + b)$ is always even. Give an example to show that he is wrong.

> Write out the first 4 or 5 prime numbers to help you find a counter-example.

....

.... *(2 marks)*

A **2** Prove that the square of any odd number is odd.

Guided

$(2n + 1)^2 = $ ....

> Use $2n$ to represent an even number and $2n + 1$ to represent an odd number. Expand and simplify $(2n + 1)^2$. You have to show that it is equivalent to one more than an even number to finish the proof.

....

....

.... *(3 marks)*

A **3** Prove that the sum of the squares of two consecutive integers is an odd number.

Guided

$n^2 + (n + 1)^2 = $ ....

....

.... *(3 marks)*

A **4** Prove that the sum of four consecutive integers is always divisible by 2.

Guided

$n + (n + 1) + (n + 2) + (n + 3) = $ ....

> Start by adding the four consecutive integers, then show that the simplified sum can be divided by 2.

....

.... *(3 marks)*

A* **5** Two integers have a difference of 4.
The difference between the squares of the two integers is four times the sum of the integers.

> If the first number is n then the other number will be $n + 4$.

For example, $10 - 6 = 4$, $10^2 - 6^2 = 100 - 36 = 64$ and $4 \times (10 + 6) = 4 \times 16 = 64$.
Prove this result algebraically.

....

....

....

.... *(4 marks)*

A* **6** Two odd numbers have a difference of 10.
Prove that the difference between their squares is equal to 10 times their sum.

....

....

....

....

.... *(4 marks)*

Identities

Guided

1 Show that $(n + 2)^2 + (n - 2)^2 \equiv 2(n^2 + 4)$

$(n + 2)(n + 2) + (n - 2)(n - 2) \equiv$ ....

....

....

.... *(2 marks)*

> Start with the expression on the left-hand side. You need to rearrange it to make the expression on the right-hand side.

2 Show that $5(x + 1)^2 - 5(x + 1) \equiv 5x(x + 1)$

....

....

....

.... *(2 marks)*

3 Given that $x^2 + ax - 2 \equiv (x - 4)^2 - b$
Work out the values of a and b.

....

....

....

....

Answer $a =$ ...., $b =$ .... *(3 marks)*

> Expand the brackets on the right-hand side and simplify. You can find a by equating the coefficients of the x terms. You can find b by equating the number terms.

4 Work out the values of a and b if $x^2 + ax + b \equiv (x - 9)^2 - a$

....

....

....

....

Answer $a =$ ...., $b =$ .... *(3 marks)*

5 Given that $a(3x + 1) + b(2x - 6) \equiv 16x - 28$
Work out the values of a and b.

> Expand the brackets. Write down two simultaneous equations and solve them to work out a and b.

....

....

....

....

....

....

Answer $a =$ ...., $b =$ .... *(5 marks)*

Completing the square

A

Guided

1 Find the values of a and b such that $x^2 + 4x = (x + a)^2 + b$

$x^2 + 4x = (x + \text{.........................})^2 - \text{..................................}$

> The identity
> $x^2 + 2bx + c \equiv (x + b)^2 - b^2 + c$
> can help here.

....

Answer.. *(2 marks)*

A

Guided

2 Find the values of a and b such that $x^2 - 6x - 1 = (x + a)^2 + b$

$x^3 - 6x - 1 = (x - \text{.........................})^2 - \text{.....................} - 1$

Answer.. *(2 marks)*

A*

3 (a) Find the values of p and q such that $x^2 - 6x + 3 = (x + p)^2 + q$

....

Answer.. *(2 marks)*

(b) Hence, or otherwise, solve the equation:
$x^2 - 6x + 3 = 0$
Give your answer in surd form.

> Put your answer to part **a** equal to 0 and solve for x.
> Remember that when you take square roots of both sides, there are two possible values. You can use $\pm$ to show both values.

....

....

....

Answer.. *(2 marks)*

A*

4 Solve the equation $x^2 - 8x + 10 = 0$
Give your answer in surd form.

....

....

....

....

....

Answer.. *(3 marks)*

A*

5 Solve the equation $x^2 - 10x + 13 = 0$
Give your answer in the form $a \pm b\sqrt{c}$ where a, b and c are integers.

....

....

....

....

....

Answer.. *(4 marks)*

57

Problem-solving practice 1

1 w, x and y are three positive integers.

w is 20% of y.

Guided x is one-sixth of y.

y is less than 100.

What values could y take?

Show clearly how you work

out your answer.

> 20% is equivalent to one-fifth. Because w and x are both **integers**, y must be a multiple of 5 **and** a multiple of 6. You need to find the numbers less than 100 which are multiples of 5 and 6.

y is a multiple of 5 and a multiple of 6 so it must be a multiple of the LCM of 5 and 6,

which is 30. The multiples of 30 less than 100 are ...., ... and ....

Check: y = ..., w = ..., x = ...

Check: y = ..., w = ..., x = ...

Check: y = ..., w = ..., x = ...

Answer .... *(4 marks)*

2 Lizzie buys 120 scarves at £2.50 each and 180 hats at £4 each to sell at the market.

She wants to sell them all and make at least 20% profit.

She sells $\frac{3}{4}$ of the scarves at £4.50 each and the rest at £2 each.

She sells $\frac{2}{3}$ of the hats at £5 each.

> Start by working out how much money Lizzie needs to make in total. Work out how much she spends on scarves and hats, then add 20%.

Lizzie can reduce the price of the remaining hats and still make at least 20% profit.

Work out the lowest price of the hats.

You **must** show your working.

....

....

....

....

....

....

....

Answer £ .... *(6 marks)*

3 Here is an algebraic addition pyramid.

Each expression is the sum of the two expressions below it.

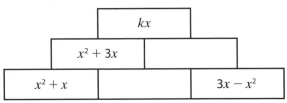

Work out the value of k.

Answer k = ... *(4 marks)*

Problem-solving practice 2

4 The diagram shows a line *ABCD*.
A is the point $(-80, 210)$

 Guided

B is the point $(140, -120)$

The line cuts the *y*-axis at *C* and the *x*-axis at *D*.

Work out the coordinates of *C* and *D*.

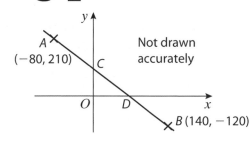

Gradient of line = ...

....

....

....

....

C = ... D = ... *(4 marks)*

5 Show that the sum of four consecutive integers must be an even number.

....

....

....

| If a number is even, it must be divisible by 2. Usually a good way to show this is to factorise the simplified expression, putting 2 outside the brackets. |

| If the first number is *n*, the next consecutive number is $n + 1$. |

....

....

.... *(4 marks)*

6 Two integers have a difference of 2.
The difference between the squares of the two integers is twice the sum of the integers.
For example, $12 - 10 = 2$, $12^2 - 10^2 = 144 - 100 = 44$
and $2 \times (12 + 10) = 2 \times 22 = 44$

| If the lowest number is *n* the other number must be $n + 2$. |

Prove this result algebraically.

....

....

....

....

....

....

....

....

Answer .... *(4 marks)*

Proportion

1   A printer takes 18 seconds to print 12 copies of a document.
How long would it take to print 40 copies of the same document?

> **Guided**

EXAM ALERT

> Students have struggled with exam questions similar to this – **be prepared!**

Time for 1 copy = $\dfrac{18 \text{ seconds}}{12}$ = ... seconds.

Time for 40 copies = 40 × .... = ... seconds.

Answer .... seconds   *(2 marks)*

2   Four gardeners usually plant a large sack of bulbs in 1 hour and 12 minutes.
One of the gardeners is off work.
How long will the three gardeners, working at the same rate, take to plant a large sack of bulbs?

> **Guided**

Total working time usually needed = 4 × 1 hour and 12 minutes

= 4 × 72 minutes = ... minutes.

Working time needed when there are three gardeners

= .... ÷ 3 minutes.

> The three gardeners would each have an equal share of the work to do.

Answer .... minutes   *(2 marks)*

3   A lorry uses 144 litres of fuel on a 320 mile journey.
How much fuel does it use on a journey of 100 miles?

....

....

....

Answer .... litres   *(2 marks)*

4   Seven people pay £16 each to hire a minibus.
How much would it cost each person if eight people hire the same minibus?

> **Guided**

Total cost = 7 × £16 = £.........................

Cost for each of the eight people = £......................... ÷ 8

> The cost must be shared equally.

= £.........................

Answer £ ....   *(2 marks)*

5   Twelve technicians install an industrial computer system in seven and a half hours.
How long would it take for 10 technicians to install the same computer system?

> Remember, first work out the total number of hours the 12 technicians took to do the job.

....

....

....

Answer .... hours   *(2 marks)*

Trial and improvement

1 Aimee is using a trial and improvement method to find a solution to this equation: $x^3 - 2x = 31$

The first trial is shown in the table.

x	$x^3 - 2x$	Comment
3	21	too low
4	56	too high
3.5	35.875	too high
3.4		

Check that when $x = 3$, you get 21 on your calculator.

As $x = 3$ is too low, try $x = 4$ next.

As x is between 3 and 4, try 3.5 next.

As $x = 3.5$ is too high, try $x = 3.4$ next.

When you have two consecutive 1 decimal place numbers, where one is too low and one is too high, you must work out the answer when x is half way between them to say which one is the closest.

Continue the trial and improvement method to find a solution to the equation.

Give your answer correct to 1 decimal place.

Answer .... *(4 marks)*

2 Use trial and improvement to find the solution to the equation: $x^3 + 3x = 400$
The first step is shown in the table.
Give your solution correct to 1 decimal place.

x	$x^3 + 3x$	Comment
6	234	too small

234 is a long way from 400, but it's still probably best to try $x = 7$ next.

Answer .... *(4 marks)*

3 Use trial and improvement to find the solution to the equation:
$x^2\sqrt{x} = 100$

Use $x = 7$ to check that you are calculating $x^2\sqrt{x}$ correctly on your calculator.

The first step is shown in the table.
Give your solution correct to 1 decimal place.

x	$x^2\sqrt{x}$	Comment
7	129.64	too high

Answer .... *(4 marks)*

The quadratic formula

Guided

1 Solve the quadratic equation $x^2 + 4x - 3 = 0$
Give your answers to 2 decimal places.
You **must** show your working.

> You have to solve a quadratic equation to 2 decimal places, so use your calculator and the quadratic formula.

$x = \dfrac{-b \pm \sqrt{b^2 - 4ac}}{2a}$ where $a = 1$, $b = 4$ and $c = -3$

> Always write down the values of a, b and c before you start.

$= \dfrac{-4 \pm \sqrt{4^2 - 4 \times 1 \times (-3)}}{2 \times 1}$

$= \dfrac{-4 \pm \sqrt{16 + 12}}{2} = \dfrac{-4 \pm \sqrt{28}}{2}$

> Now you have to work out $x = \dfrac{-4 + \sqrt{28}}{2}$ and $x = \dfrac{-4 - \sqrt{28}}{2}$, giving both answers to 2 decimal places.

....

....

Answer .... *(3 marks)*

2 Solve the equation $2x^2 - 4x - 4 = 0$
Give your answers to 2 decimal places.
You **must** show your working.

> Remember to write down the quadratic formula and the values of a, b and c before you start. Be extra careful here – there are two negative numbers.

....

....

....

....

Answer .... *(3 marks)*

3 Solve the quadratic equation

$5x^2 + 2x - 5 = 0$

Give your answers to 2 decimal places. You **must** show your working.

....

....

....

....

Answer .... *(3 marks)*

4 Solve $(3x + 2)(x - 1) = 6 - 5x$
Give your answers to 2 decimal places.
You **must** show your working.

....

....

....

....

....

Answer .... *(4 marks)*

Quadratic graphs

1 **(a)** Complete the table of values for $y = 2x^2 - 5$

x	-2	-1	0	1	2
y	3			-3	

> Check using $x = -2$. Enter $2 \times (-2)^2 - 5$ into your calculator. The answer should be 3.

(2 marks)

(b) Draw the graph of $y = 2x^2 - 3$ for values of x from -2 to 2.

> Plot each of the five pairs of coordinates in the table and join them up with a **smooth** curve.

(2 marks)

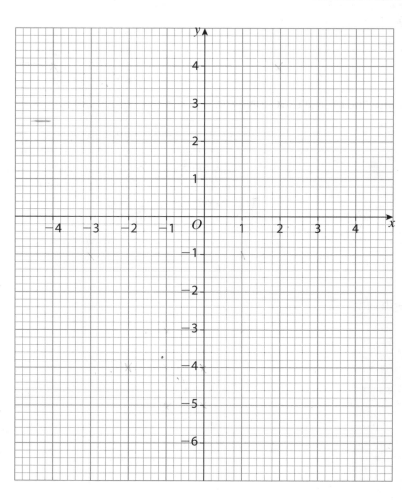

> Draw a line from $x = 1.5$ vertically to the curve you have just drawn. Write down the corresponding y value.

(c) Use the graph to find the value of y when $x = 1.5$

.... *(1 mark)*

2 **(a)** Complete the table of values for the graph of $y = x^2 + 2x - 4$

x	-4	-3	-2	-1	0	1	2
y		-1	-4		-4		4

(b) Draw the graph of $y = x^2 + 2x - 4$ for values of x from -4 to 2 on the grid above. *(2 marks)*

(c) Where does the graph of $y = x^2 + 2x - 4$ cross the x-axis?

$x =$ .... and $x =$ .... *(1 mark)*

Using quadratic graphs

1 Here is a quadratic graph of $y = x^2 - 4x + 2$

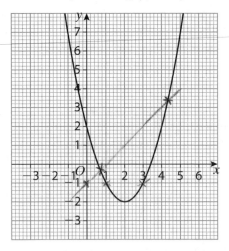

By drawing an appropriate line on the grid, solve the equation $x^2 - 5x + 3 = 0$

$x^2 - 5x + 3 = 0$ $(+ x)$

$x^2 - 4x + 3 = x$ $(- ....)$

$x^2 - 4x + 2 = x - ....$

So the line is $y = x - ....$

> Rearrange the equation to make the left-hand side equal to $x^2 - 4x + 2$. The right-hand side tells you which line to draw. The solutions will be the x-coordinates at the points where the line crosses the curve.

....

Answer ....   *(3 marks)*

2 Here is a quadratic graph of $y = 2x^2 - 2x + 3$

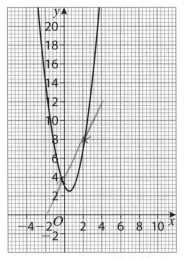

By drawing an appropriate line on the grid, solve the equation $2x^2 - 4x - 1 = 0$

$2x^2 - 4x - 1 = 0$ $(+ 2x)$

$2x^2 - 2x - 1 = 2x$ $(....^{+ 4}....)$

....

....

....

Answer ....   *(3 marks)*

Graphs of curves

1 Sketch the graph of $y = 2^x$.
Label any points of intersection with the axes.

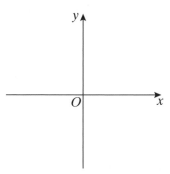

> If you don't know what the key points are, try a few values for x on your calculator. Make sure you try positive and negative numbers, and most importantly zero, as it will give the intersection with the y-axis.

(2 marks)

2 Here is the graph of $y = 3^x$.

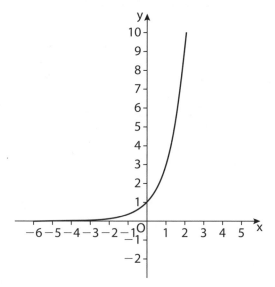

> This is a transformation from $y = f(x)$ to $y = f(-x)$. If you don't know what this looks like, try a few values of x on your calculator.

On the same set of axes, sketch the graph of $y = 3^{-x}$. *(2 marks)*

3 For each equation below, write down the letter of the graph which matches the equation.

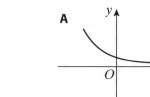
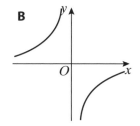
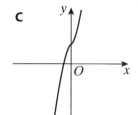
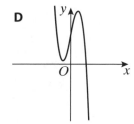

(a) $y = -\dfrac{5}{x}$

Answer .... *(1 mark)*

(b) $y = x^3 + 5x + 5$

Answer .... *(1 mark)*

(c) $y = 0.9^x$

Answer .... *(1 mark)*

3-D coordinates

C **1** *OABCDEFG* is a cuboid as shown.

Guided

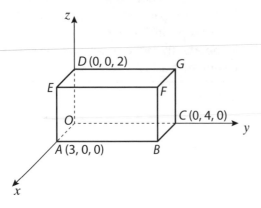

(a) Write down the coordinates of *B*.

> *B* has the same *x*-coordinate as *A* and the same *y*-coordinate as *C*. It has the same *z*-coordinates as *A* and *C*.

Answer (3, 4, ...) *(1 mark)*

(b) Write down the coordinates of *F*.

> Find the mean of each pair of coordinates. The two *x*-coordinates are 3 and 0, so the mean is 1.5.

Answer (........., ..., ...) *(1 mark)*

(c) Write down the coordinates of the midpoint of *AD*.

Answer (1.5, ..., ...) *(1 mark)*

C **2** *OABCDEFG* is a cuboid as shown.

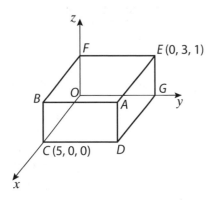

(a) Write down the coordinates of *D*.

Answer (........., ..., ...) *(2 marks)*

(b) Write down the coordinates of the midpoint of *CE*.

> Working out the coordinates of *G* first might help.

Answer (........., ..., ...) *(2 marks)*

Proportionality formulae

1 y is inversely proportional to the square of x.

When $x = 2$, $y = 5$

(a) Work out an equation connecting y and x.

> Always start by writing down the statement and then the formula.

$y \propto \dfrac{1}{x^2}$

$y = \dfrac{k}{x^2}$

> Now substitute x and y for the values given and work out the value of k.

....

....

$k = \ldots\ldots\ldots$

So, $k = \dfrac{\ldots\ldots}{x^2}$ Answer .... *(3 marks)*

(b) Work out the value of y when $x = 12$

Give your answer as a fraction in its simplest form.

> Substitute $x = 12$ into your equation.

....

....

....

Answer .... *(2 marks)*

2 b is directly proportional to $\sqrt{a}$.

When $a = 9$, $b = 15$

Calculate the value of b when $a = 2\frac{1}{4}$

> Remember: statement, formula, substitution, equation. Once that is done, you can answer the question!

....

....

....

....

....

....

Answer .... *(5 marks)*

3 t is inversely proportional to the cube of d. When $d = 2$, $t = 24$

Calculate the value of t when $d = 4$

....

....

....

....

....

....

Answer .... *(3 marks)*

Transformations 1

A* **1** **(a)** The graph of $y = 2x^2$ is transformed by the vector $\begin{pmatrix} 0 \\ 3 \end{pmatrix}$.

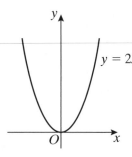   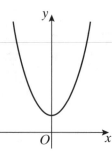

Write down the equation of the transformed graph.

> The graph has not changed shape, it's just moved up the y-axis.

Answer .... *(1 mark)*

(b) The diagram shows the graph of $y = x^2$

On the same diagram, sketch the graph of $y = (x + 1)^2$

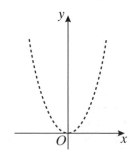

> If you don't **know** what happens here, you could substitute a few values of x into both $y = (x + 1)^2$ and $y = x^2$.

> The graph of $y = f(x - a)$ is translated by the vector $\begin{pmatrix} a \\ 0 \end{pmatrix}$.

(1 mark)

A* **2** The four diagrams all show the function $y = f(x)$

A 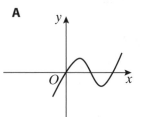   **B** 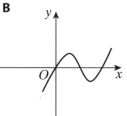   **C** 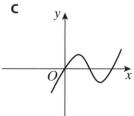   **D**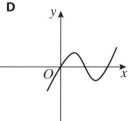

(a) On graph A sketch the function $y = f(x) + 2$ *(1 mark)*

(b) On graph B sketch the function $y = f(x + 2)$ *(1 mark)*

(c) On graph C sketch the function $y = 2f(x)$ *(1 mark)*

(d) On graph D sketch the function $y = f(2x)$ *(1 mark)*

> $y = f(x) + a$ is translated with vector $\begin{pmatrix} 0 \\ a \end{pmatrix}$.

> $y = af(x)$ is a vertical stretch, scale factor a. The x-coordinate of each point stays the same.

Transformations 2

1 Six identical graphs of $y = \sin x$ are shown for $0° \leqslant x \leqslant 360°$.

A
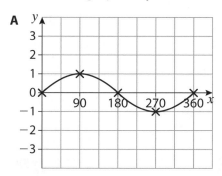

B
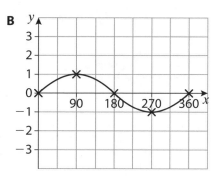

C
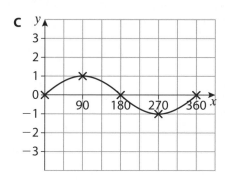

D
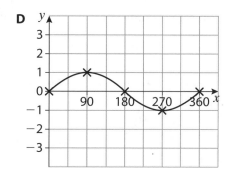

E
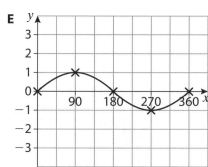

F
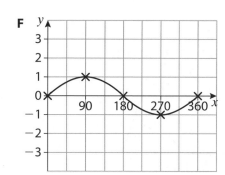

(a) On graph A draw the graph of $y = 3 \sin x$ *(1 mark)*

(b) On graph B draw the graph of $y = \sin \frac{1}{2} x$ *(1 mark)*

(c) On graph C draw the graph of $y = -\sin x$ *(1 mark)*

(d) On graph D draw the graph of $y = -\sin(-x)$ *(1 mark)*

(e) On graph E draw the graph of $y = \sin(x + 45)$ *(1 mark)*

(f) On graph F draw the graph of $y = \sin x + 2$ *(1 mark)*

> $f(x) \rightarrow af(x)$ is a vertical stretch, scale factor a.

> $f(x) \rightarrow -f(x)$ is a reflection in the x-axis, whereas $f(x) \rightarrow f(-x)$ is a reflection in the y-axis.

> $f(x) \rightarrow f(x + a)$ is a translation by the vector $\begin{pmatrix} -a \\ 0 \end{pmatrix}$.

Angle properties

 1   *AB* and *CD* are parallel.

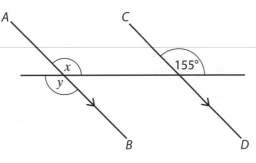

(a) Write down the value of x.
Give a reason for your answer.

Decide whether the angle property you need is corresponding, alternate or allied angles. You must learn which is which!

Answer .... degrees

Reason ....   *(2 marks)*

(b) Write down the value of y.
Give a reason for your answer.

Answer .... degrees

Reason ....   *(2 marks)*

 2   Here is a diagram of a trapezium.

> **Guided**

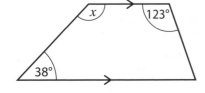

The shape is a trapezium, so the top and bottom lines are parallel.

Write down the size of angle x.
Give a reason for your answer.

$x + 38° = 180°$ ....

Answer .... degrees

Reason ....   *(2 marks)*

 3   Here is a diagram of a triangle.

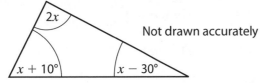   Not drawn accurately

EXAM ALERT

Students have struggled with exam questions similar to this – **be prepared!**

Set up and solve an equation to work out the value of x.

....

....

....

Answer .... degrees   *(4 marks)*

Solving angle problems

1 AB is parallel to CD.
LMN and PMQ are straight lines.
MQ = MN

> **Guided**

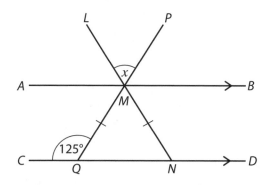

MQ = MN so the triangle MNQ is isosceles.

Remember to show your working, and give a reason for each angle that you calculate.

Work out the value of x. Give reasons for each step of your working.

Angle MQN = 180° − 125° = 55° (Angles on a straight line)

Angle MNQ = 55° (Base angles of an isosceles triangle)

....

....

Answer .... degrees   *(3 marks)*

2 CE is parallel to AF.
AB, CD and EF are parallel.
ABD = 20° and EFD = 110°

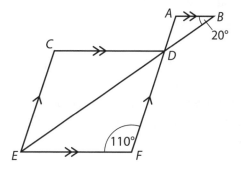

Work out each of the following, giving reasons.

(a) Angle BAF ....

Answer .... degrees

Reason ....   *(1 mark)*

(b) Angle BDA ....

Answer .... degrees

Reason ....   *(1 mark)*

(c) Angle FDE ....

Answer .... degrees

Reason ....   *(1 mark)*

(d) Angle CED ....

Answer .... degrees

Reason ....   *(1 mark)*

Angles in polygons

D **1** Show that the interior angle of a regular hexagon is 120°.

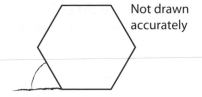
Not drawn accurately

Guided

> It's usually easier to work out the exterior angle first and use it to calculate the interior angle.
> Exterior angle + interior angle = 180°

Exterior angle = $\dfrac{360°}{n}$ = $\dfrac{360°}{6}$ = ...

....

.... *(2 marks)*

D **2** The diagram shows a regular pentagon divided into five congruent triangles.

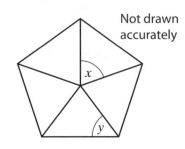
Not drawn accurately

> The angles around a point add up to 360°.

(a) Work out the value of x.

....

....

Answer .... degrees   *(2 marks)*

(b) Work out the value of y.

> Each triangle is isosceles.

....

Answer .... degrees   *(2 marks)*

C **3** The diagram shows a tile in the shape of an isosceles trapezium. Some of these tiles are put together to make a regular shape. The diagram below is incomplete.

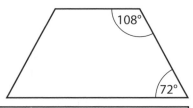
108°
72°

> Students have struggled with exam questions similar to this – **be prepared!**

How many exterior sides does the shape have?
You **must** show your working.

....

....

....

....

....

....

Answer .... sides   *(3 marks)*

Circle facts

1 The diagram shows a circle, centre *O*.
AT and *BT* are tangents to the circle.
Angle *OAB* = 35°.

> **Guided**

> Remember that you must say which circle facts you are using.

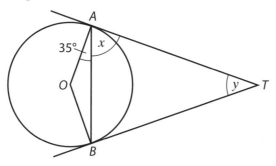

(a) Work out the value of *x*.

OA is a radius and *AT* is a tangent.

The angle between a radius and a tangent = ... degrees

> If you are not sure what to do – find as many angles as you can and eventually you will work out angle *x*. Make sure you show workings and give reasons.

....

....

Answer .... degrees   *(1 mark)*

(b) Work out the value of *y*.

Triangle *ABT* is isosceles.

....

....

Answer .... degrees   *(2 marks)*

2 The diagram shows a circle, centre *O*.
PQ and *RQ* are tangents to the circle.
Angle *POR* = 236°
Work out the value of angle *OQR*.

> The line segment *OQ* bisects the quadrilateral *OPQR*, forming two congruent triangles.

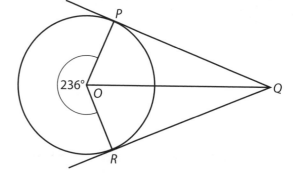

....

....

....

....

....

....

Answer .... degrees   *(4 marks)*

Circle theorems

B-A **1** The six diagrams below show circles, centre *O*. Work out the value of *x* in each case.

> Take your time to decide which circle theorem you need to use. Draw on the diagrams if it helps you.

(a)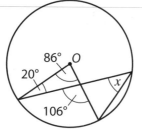

x = ... degrees
(1 mark)

(b)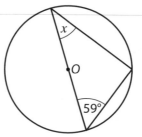

x = ... degrees
(2 marks)

(c)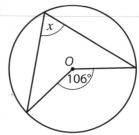

x = ... degrees
(1 mark)

(d)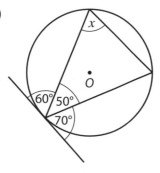

x = ... degrees
(1 mark)

(e)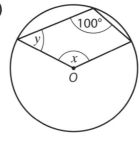

x = ... degrees
(2 marks)

(f)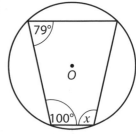

x = ... degrees
(1 mark)

A* **2** *PQRS* is a cyclic quadrilateral. *AB* is a tangent to the circle at *S*.

 Guided

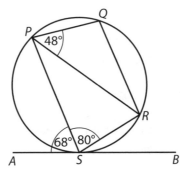

Prove that *PS* is parallel to *QR*.

Angle PRS = 68° (alternate segment theorem)

> It is easiest to show that *SPR* and *PRQ* are alternate angles.

....

....

....

....

....

....

.... *(5 marks)*

Perimeter and area

D 1 The diagram shows a scale drawing of a room.

(a) Work out the perimeter of the room.

| Work out the two missing lengths first. |

....

....

....

Answer .... metres   *(1 mark)*

(b) Work out the floor area of the room.

| Split the compound shape into two rectangles before working out the area for each. |

....

....

....

Answer .... m²   *(2 marks)*

D 2 The diagram shows a flower bed. It is in the shape of a parallelogram and a right-angled triangle.
A gardener wants to cover the flower bed with a layer of compost.
A jumbo-bag of compost covers 8.5 m² of the flower bed.
How many jumbo-bags of compost are needed?

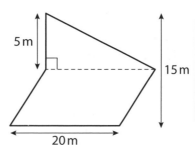

| Work out the area of the parallelogram and the triangle separately. When you have worked out the total area divide by 8.5 and remember to round your answer sensibly. |

....

....

....

....

Answer .... bags   *(4 marks)*

D 3 Work out the area of the shaded part in this diagram.

Guided

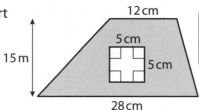

| Take the area of the square away from the area of the trapezium. |

Area of trapezium = ½(a + b) h

         = ½ (......... + ...) × .... = ...

Area of square = ...

| The formula for the area of a trapezium is given on the formula sheet. |

....

....

....

Answer .... cm²   *(4 marks)*

Similar shapes 1

B

1 *ABC* and *PQR* are similar triangles.
Work out the length of *QR*.

Guided

$$\frac{QR}{BC} = \frac{PQ}{AB}$$

$$\frac{QR}{18} = \frac{4.8}{24}$$

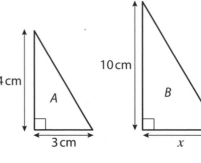

Not drawn
accurately

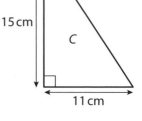

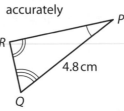

You know *AB*, *BC*
and *PQ* and you
need to find *QR*.

QR = ...

Answer .... cm   *(3 marks)*

B

2 *A*, *B* and *C* are right-angled
triangles.
A and *B* are similar.

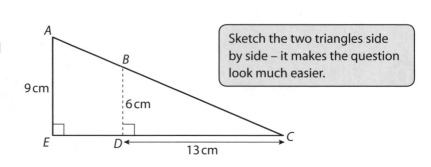

Not drawn accurately

(a) Work out the length *x* on triangle *B*.

....

....

....

Answer .... cm   *(3 marks)*

(b) Show that triangles *A* and *C* **are not** similar.

....

....

Answer ....   *(2 marks)*

B

3 The diagram shows a triangle
cut into a smaller triangle and
a trapezium.

Sketch the two triangles side
by side – it makes the question
look much easier.

Work out the length *DE*.

....

....

....

....

Answer .... cm   *(4 marks)*

Congruent triangle proof

B **1** Each pair of triangles are congruent.
State the condition that shows the triangles are congruent.

Mark on the diagrams which angles or sides are equal on both triangles.

(a)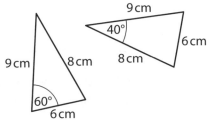

(b)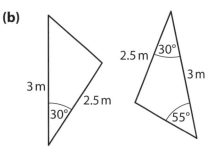

(c)

Condition:.....................

....

Condition:.....................

....

Condition:.....................

....

(3 marks)

A **2** *ABCD* is a parallelogram.

Remember to give reasons for any statements you make using the properties of a parallelogram and the angle properties of parallel lines.

Prove that triangle *ABC* is congruent to triangle *ACD*.

AB = DC and BC = AD (opposite sides of a parallelogram are equal)

....

....

....

....

.... *(4 marks)*

A* **3** The diagram shows two squares, *PQRS* and *STUV*.
Use congruent triangles to prove angle *PTS* = angle *RVS*.

Students have struggled with exam questions similar to this – **be prepared!**

Use triangles *PST* and *RSV*.

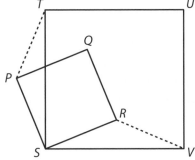

....

....

....

....

.... *(4 marks)*

Pythagoras' theorem

C

Guided

1 Work out the length of *AC*.

Using Pythgoras' theorem $a^2 + b^2 = c^2$

$$35^2 + 25^2 = AC^2$$

$$1225 + 625 = AC^2$$

$$\text{.........................} = AC^2$$

$$AC = \sqrt{\text{.........................}}$$

$$AC = \text{.........................}$$

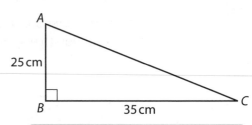

Write down your whole calculator display before rounding your answer to a suitable degree of acuracy.

Answer .... cm   *(3 marks)*

C

Guided

2 The screen size of a television is the length, to the nearest inch, of the diagonal of the screen.

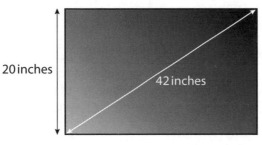

Sketch and label a right-angled triangle and write on it the two dimensions given.

The screen on this 42 inch television is 20 inches high.
How wide is the screen? Give your answer to the nearest inch.

Using Pythgoras' theorem $a^2 + b^2 = c^2$

$$BC^2 + 20^2 = 42^2$$

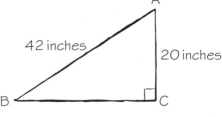

....

....

....

Answer .... inches   *(3 marks)*

C

3 Two mobile phone masts are 2.1 km apart on horizontal ground. Work out the distance between the tops of the masts. Give your answer to the nearest metre.

Sketch and label a right-angled triangle.

....

....

....

....

....

....

Answer .... m   *(4 marks)*

Pythagoras in 3-D

1 The diagram shows a cuboid *ABCDEFGH*. Work out the length *AH*.

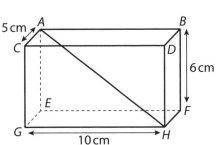

> Either use the formula $a^2 + b^2 + c^2 = d^2$, or if you prefer to use Pythagoras' theorem twice, work out *EH* first.

....

....

....

....

....

Answer ....   *(3 marks)*

2 The diagram shows a wedge *ABCDEF*. Work out the length *BE*.

> Use the same method you used in Q1.

....

....

....

....

....

Answer ....   *(3 marks)*

3 The diagram shows a pyramid *ABCDE*. The base *ABCD* is a square of side length 10 cm. The length of each sloping edge is 20 cm. Work out the vertical height of the pyramid.

> Work out *AC* (or *BD*) first, then sketch another right-angled triangle involving the vertical height and the sloping edge.

....

....

....

....

....

Answer .... cm   *(5 marks)*

Trigonometry 1

B

EXAM ALERT

2 Calculate the size of angle x.

hyp
12 cm

5 cm
opp

Not drawn accurately

adj

x

S^O_H　C^A_H　T^O_A

> Label the sides first, then decide which ratio to use.

> You need to use the $\sin^{-1}$ function button on your calculator.

> Students have struggled with exam questions similar to this – **be prepared!**

....

....

....

....

Answer .... degrees　*(3 marks)*

B

2 A ladder 5 m long rests against a wall. The foot of the ladder is 2.1 m from the base of the wall.
What angle does the ladder make with the floor?
Give your answer to 3 significant figures.

5 m

2.1 m

> Label the sides of the right-angled triangle first.

....

....

....

....

....

Answer .... degrees　*(3 marks)*

B

3 A passenger in an aircraft is told by the pilot that they are flying at an altitude of 3000 m and are 8 km horizontally from the airport.
Someone at the airport is looking up at the plane.
At what angle to the horizontal are they looking up at the plane?

plane

3000 m

airport

8 km

> Label the sides of the right-angled triangle and take care with units.

....

....

....

....

Answer .... degrees　*(3 marks)*

Trigonometry 2

B **1** Work out the length of side x.

Guided

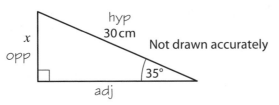

Not drawn accurately

Label the sides first, then decide which ratio to use.

S^O_H C^A_H T^O_A

....

....

....

....

Remember to check that your answer looks sensible.

Answer .... degrees   *(3 marks)*

B **2** Passengers in an aircraft are told by the pilot that they are flying at an altitude of 5000 m. Someone at the airport is looking up at the plane at an angle of elevation of 40°. What is the horizontal distance of the plane from the airport?

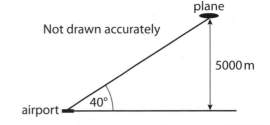

Not drawn accurately

....

....

Label the sides of the right-angled triangle first, then decide whether to use S^O_H C^A_H or T^O_A.

....

....

....

Answer .... km   *(3 marks)*

B **3** The diagram shows a square. The diagram has one line of symmetry. Calculate the area of the shaded part of the square.

Not drawn accurately

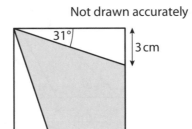

Use trigonometry to work out the length of the side of the square. With this measurement you can work out the area of the square and each of the triangles.

....

....

....

....

....

Answer .... cm²   *(5 marks)*

The sine rule

1 Work out the size of angle x.

 Guided

$$\frac{\sin A}{a} = \frac{\sin B}{b}$$

EXAM ALERT

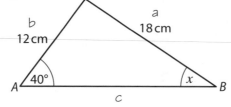

Substitute in the given values and rearrange for sin x, then use the $\sin^{-1}$ function to work out x.

Students have struggled with exam questions similar to this – **be prepared!**

....

....

....

....

Answer .... degrees   *(3 marks)*

2 The diagram shows triangle *PQR*. Work out the length of side *PR*.

Guided

$$\frac{PR}{\sin Q} = \frac{QR}{\sin P}$$

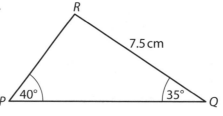

....

....

....

....

Answer .... cm   *(3 marks)*

3 The diagram shows a vertical mast, held in place by two wires. To work out the length of the longest wire, Hamish has made some measurements as shown on the diagram.

Guided

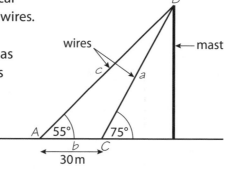

Label the vertices of the triangle *ABC* and the sides *abc*. Notice that the 75° angle is not within the triangle you are using. Write on the diagram the size of angle *ACB*.

Work out the length of the longest wire.

....

....

....

....

....

Answer .... m   *(4 marks)*

The cosine rule

A **1** The diagram shows a triangle. Work out the length of the longest side of the triangle.

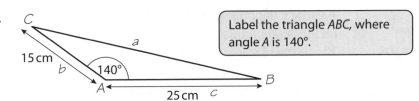

Label the triangle *ABC*, where angle *A* is 140°.

Guided

$$a^2 = b^2 + c^2 - 2bc \cos A$$

....

....

....

....

This equation is given on the formula sheet.

Answer .... cm   *(3 marks)*

A **2** The diagram shows triangle *ABC*. Show that angle *ACB* is 33.3°.

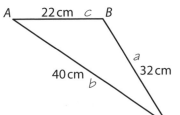

$c^2 = a^2 + b^2 - 2ab \cos C$ is a rearrangement of the equation given on the formula sheet.

Guided

$$c^2 = a^2 + b^2 - 2ab \cos C$$

$$\cos C = \frac{a^2 + b^2 - c^2}{2ab}$$

Use this version of the cosine rule to find a missing angle.

....

....

....

....

....   *(3 marks)*

A* **3** Here is a quadrilateral *ABCD*. Work out the size of angle *BDC*.

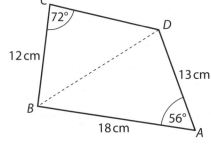

....

....

....

....

....

....

Answer .... degrees   *(6 marks)*

Trigonometry in 3-D

A*

Guided

1 A vertical flagpole *AE* stands at the corner of a rectangular parade ground *ABCD*.
The angle of elevation of the top of the flagpole from *C* is 8°.
Work out the height of the flagpole.

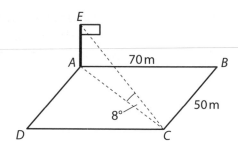

First work out the diagonal *AC* using the triangle *ABC*.

Using Pythagoras' theorem $AC^2 = 70^2 + 50^2$ ....

....

....

Sketch triangle *ACE*, write on it the information you have, label the sides hyp, opp and adj and decide whether to use S_H^O, C_H^A or T_A^O.

....

....

....

Answer .... m   *(4 marks)*

A*

2 In this diagram, *CD* = 50 cm, angle *BDC* = 40° and angle *ACB* = 20°
Angles *ABC*, *ABD* and *BCD* are all 90°

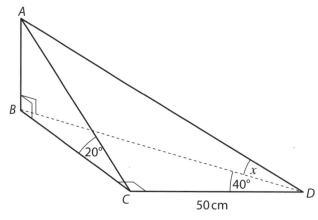

Work out the size of angle *x*.
Give your answer correct to 3 s.f.

....

....

....

....

....

....

....

....

Answer .... degrees   *(6 marks)*

Circles and cylinders

1 The circle is drawn accurately.
Work out the area of the circle.
Give your answer in terms of π.
State the units of your answer.

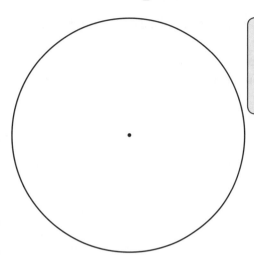

> Measure the radius very carefully. Write down the formula you will use to work out the area of the circle.

....

....

....

....

....

....

....

Answer .... *(3 marks)*

2 The diagram shows a semi-circular shape.

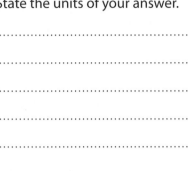

Not drawn accurately

> Students have struggled with exam questions similar to this – **be prepared!**

r cm

Circle the correct expression for the perimeter of the shape.

$\frac{1}{2}\pi r^2$ $2\pi r$ $\frac{1}{2}\pi r^2 + 2r$ $\pi r + 2r$ $\frac{1}{2}\pi r$

.... *(1 mark)*

3 The diagram shows a cylinder.
The diameter of the base is 14 cm.
The height is 12 cm.

> Guided

Work out the volume of the cylinder.
Give your answer to a suitable
degree of accuracy.

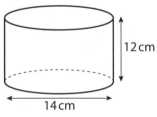

12 cm

14 cm

> Write down the formula for the volume of a cylinder from memory – it's not on the formula sheet!

> Work out the radius then substitute r and h into the formula.

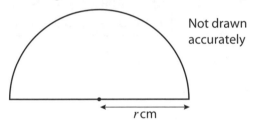

Volume of cylinder $= \pi r^2 h$

....

....

....

....

....

> Write down all the digits off your calculator display then round to a suitable degree of accuracy.

Answer .... cm³   *(3 marks)*

Sectors of circles

A

1 A major sector is cut out of a circle of radius 6 cm.

> **Guided**

(a) Work out the perimeter of the sector.
Give your answer correct to 1 decimal place.

$$\text{Arc length} = \frac{x}{360°} \times 2\pi r$$

$$= \frac{.........}{360°} \times 2 \times \pi \times ....$$

$$= .... \text{ cm}$$

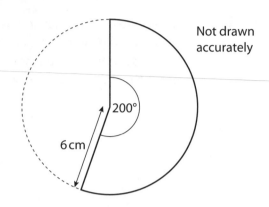

Not drawn accurately

200°

6 cm

....

....

....

> The most common mistake is not to add the two radii to the arc length.

Answer .... cm   *(4 marks)*

(b) What area of the original circle is left?
Give your answer in terms of π.

> The remaining area is the **minor** sector of the circle. Its angle is **not** 200°.

....

....

....

Answer ....   *(3 marks)*

A*

2 The diagram shows a major sector of a circle of radius 8 cm and area 151 cm². Work out angle x correct to 3 significant figures.

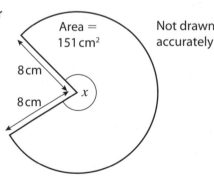

Area = 151 cm²

Not drawn accurately

8 cm

8 cm

x

> Write out the formula for the area of a sector, substitute in all given values and rearrange for x.

....

....

....

....

....

....

....

....

Answer .... degrees   *(3 marks)*

Triangles and segments

1   ABC is a triangular plot of land.
This land is valued at £15 000 per hectare.

Guided

One hectare is 10 000 m².
Work out the value of the plot ABC.

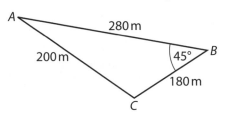

> You don't need to use all the information on the diagram – be careful to choose the correct two lengths.

Area of triangle $ABC = \frac{1}{2}ab \sin C$

....

....

....

Answer £........................   *(4 marks)*

A*   **2**   The diagram shows a shaded
segment of a circle.
The circle has a radius of 20 cm.
The angle in the minor sector is 125°.
Work out the area of the shaded
segment.

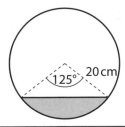

> Area of segment = area of whole sector − area of triangle.

....

....

....

....

....

Answer .... cm²   *(5 marks)*

A*   **3**   The diagram shows a regular hexagon which fits exactly inside
a circle of radius 10 cm.
Work out the shaded area between the circle and the hexagon.

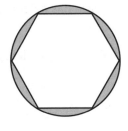

....

....

....

....

....

Answer .... cm²   *(6 marks)*

Prisms

1 The diagram shows a cuboid.

15 cm   1.3 m
75 cm

(a) Work out the volume of the cuboid.

Volume of cuboid = l × w × h

= .... × .... × ....

> Notice that the units are not the same. Convert 1.3 m into centimetres.

Answer .... cm³   *(2 marks)*

(b) Work out the surface area of the cuboid.

Surface area = 2 × l × w + 2 × l × h + 2 × w × h

= ....

> Add the areas of all six faces to work out the total surface area.

Answer .... cm²   *(2 marks)*

2 The diagram shows a warehouse. The four outside walls of the warehouse need painting. The roof does **not** need to be painted.

One can of paint covers 25 m². How many cans of paint are needed?

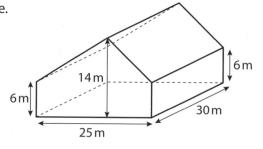

14 m
6 m
6 m
30 m
25 m

> To work out the area of the end wall split it into a triangle and a rectangle.

....

....

....

....

....

Answer .... cans of paint   *(4 marks)*

3 Lissie has a cube of clay, side length 6 cm. She remodels the clay into a cuboid.

Work out the length, x, of the cuboid.

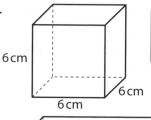

6 cm
6 cm
6 cm
2 cm
2 cm
x

> The two volumes are the same, so 2 × 2 × x is the same volume as 6 × 6 × 6.

....

....

....

....

....

....

Answer .... cm   *(4 marks)*

Volumes of 3-D shapes

1 Cylinder B has twice the volume of cylinder A.
Work out the height, *h*, of cylinder B.

18 cm A 20 cm   B *h* cm 15 cm   Not drawn accurately

....

....

....

Answer .... cm   *(4 marks)*

2 The diagram shows a rectangular-based pyramid. Work out the volume of the pyramid.

> The volume of a pyramid is not given on the formula sheet – you must learn it.

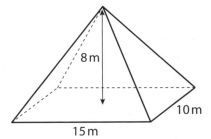

8 m   10 m   15 m

Volume of pyramid = $\frac{1}{3}$ × base area × vertical height

....

....

Answer .... m³   *(3 marks)*

3 Flynn is making spheres out of clay.
A box of clay contains 20 packs.
Each pack is a cuboid measuring
10 cm by 10 cm by 4 cm.

> Work out the total volume of clay in a box. Work out the volume of one sphere – the formula is on the formula sheet, it is $\frac{4}{3}\pi r^3$. Then divide to work out how many spheres can be made.

(a) How many spheres of radius 5 cm can Flynn make from a box of clay?

....

....

....

....

Answer .... spheres   *(4 marks)*

Each sphere of radius 5 cm is packed into a cubical box with side length 10 cm.

(b) How much spare space does the box contain?   *(2 marks)*

....

....

Answer .... cm³   *(2 marks)*

Surface area

A

〉Guided〉

1 The diagram shows a solid in the shape of a cone.
Work out the total surface area of the solid.

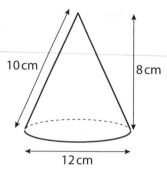

10 cm   8 cm

12 cm

> You have been asked for the total surface area. You must find the total of the curved surface area and the flat circular base area.

Total surface area = πrl + ...

= $\pi \times$ .... $\times$ .... + ...

....

Answer .... cm² *(4 marks)*

A

2 The diagram shows a litter bin made from a cylinder and a hemisphere.
The cylinder is 1 metre tall and has a diameter of 60 cm.
Work out the curved surface area of the outside of the litter bin.

....

....

....

....

....

....

....

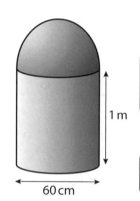

1 m

60 cm

> The radius of the hemisphere is the same as the radius of the cylinder.

> Do not include the base of the litter bin.

Answer .... m² *(4 marks)*

A

3 A sphere of radius 9 cm has the same surface area as a hemisphere.
Work out the radius of the hemisphere.

Give your answer correct to 1 decimal place.

....

....

$r = 9$ cm $r = ?$ cm

....

....

....

....

Answer .... cm *(5 marks)*

Plans and elevations

D   **1**   The diagram shows a shape made with 8 cubes.

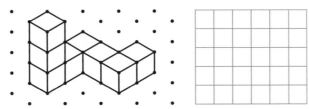

On the square grid paper, draw the plan view of the shape.

> The plan view is the view looking straight down from directly above the shape.

(1 mark)

D   **2**   The diagram shows the front elevation, side elevation and plan view of a shape made from **8** cubes.

Plan   Side elevation   Front elevation

Draw the 3-D shape on the isometric grid paper below.

> Use a ruler to draw your lines, and make sure that they always go between the dots on the isometric paper. Your finished shape should use 8 cubes.

(3 marks)

Bearings

1 The diagram shows the positions of four friends.

Choose the correct bearing from the list to complete each sentence. *(4 marks)*

040°   055°   120°   140°   220°   240°   305°   355°

> For parts **(a)** and **(b)** draw a north line from Chen. Drawing lines from Chen to Abu and Julia will help you to measure accurately. Remember that bearings are always measured from north in a clockwise direction.

(a) Abu is on a bearing of .... from Chen.   *(1 mark)*

(b) Julia is on a bearing of .... from Chen.   *(1 mark)*

(c) Abu is on a bearing of .... from Toni.   *(1 mark)*

(d) Julia is on a bearing of .... from Toni.   *(1 mark)*

2 Four footpaths meet at *O*.
C is due South of *O*.

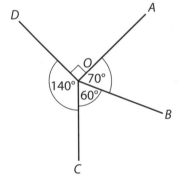

> Students have struggled with exam questions similar to this –
> **be prepared!**

> Draw a line from *O* due north. Use angle facts to work out the bearing you need.

(a) Priah is at *O* and walks towards *A*.
On what bearing does she walk?

....

Answer .... degrees   *(2 marks)*

(b) Alan is at *A* and walks to meet Priah.
On what bearing does he walk?

....

Answer .... degrees   *(2 marks)*

(c) Bim is walking from *D* to *O*.
When he gets to *O* what angle **clockwise** does he need to turn through to continue to *C*?

> Extending the line *DO* will help.

....

Answer .... degrees   *(2 marks)*

Scale drawings and maps

1 The diagram shows the map of an island.
The scale of the map is 1 : 50 000.
Calculate the actual distance from
West Hill to East Point in kilometres.

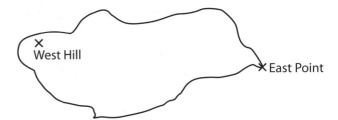

Distance in cm = ... cm × 50 000

          = ... cm

.... cm ÷ 100 = ... m

.... m ÷ 1000 = ... km

Measure the distance between the two crosses
in cm. Use the scale to calculate the actual distance
in centimetres between them, then convert your
answer into kilometres.

Answer .... km   *(3 marks)*

2 The drawing shows a scale model of a building. On the model, the building is 32.5 cm long.

Scale = 1 : 2000

32.5 cm

Work out the actual length of the building in metres.

....

....

Answer .... m   *(2 marks)*

3 The diagram shows a scale drawing of a port and two lighthouses.

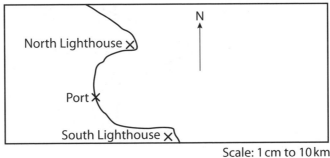

Scale: 1 cm to 10 km

A boat leaves the port and sails 50 km on a bearing of 075°.

(a) Mark the position of the boat with a cross.

Measure to the nearest mm and degree.

(2 marks)

(b) Which lighthouse is the boat
closest to, and by how many
kilometres?

Show working for the distance of the boat to both
lighthouses before saying which is closest and by
how many kilometres.

....

....

....

....

Answer .... by .... km   *(5 marks)*

Constructions

D **1** Use ruler and compasses only to make an accurate copy of triangle *ABC*.
Show all construction lines and arcs.
Line *AB* has been drawn for you.

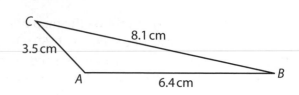

A   6.4 cm   B
3.5 cm   C   8.1 cm

> Use your compasses to draw arcs from *A* and *B*. The point where they cross is *C*. Check your finished triangle by measuring.

A ————————————— B
  6.4 cm

(2 marks)

C **2** Use ruler and compasses only.

(a) Construct the line perpendicular to *AB* that passes through point *P*.

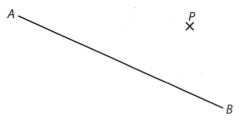

> Start by placing your compass point on *P* and drawing two arcs with the same radius which intersect *AB*.

(2 marks)

(b) Construct the angle bisector of the angle *XYZ*.   *(2 marks)*

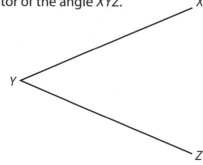

C **3** The diagram shows a scale drawing of a room. The dimensions of the room are 7 m by 4 m.
Two plug sockets are fitted along the walls at points *A* and *B*.
A third plug socket is to be fitted in wall *XY*.
The third socket must be equidistant from *A* and *B*.

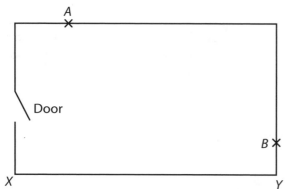

Scale: 1 cm represents 1 m

> Start by constructing the perpendicular bisector of *AB*.

Using ruler and compasses, show the position of the third socket. Label it *C*.   *(2 marks)*

Loci

 1 Two radio stations at *A* and *B* pick up a distress call from a boat at sea.
The station at *A* can tell that the boat is between 50 km and 70 km from *A*.
The station at *B* can tell that the boat is on a bearing of between 050° and 070° from *B*.

Scale: 1 cm represents 10 km

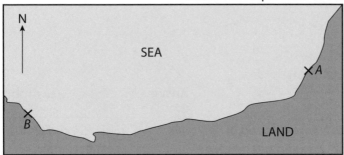

> You will need to draw two circles with their centres at point *A* and two lines which pass through point *B*.

Show clearly, using compasses and a protractor, the region where the boat will be found.

(3 marks)

 2 The diagram shows a scale drawing of a field and a shed. A goat is tethered to a rope 7 m long.
The other end of the rope is fixed to point *A* at the corner of the shed.

Scale: 1 cm represents 1 m

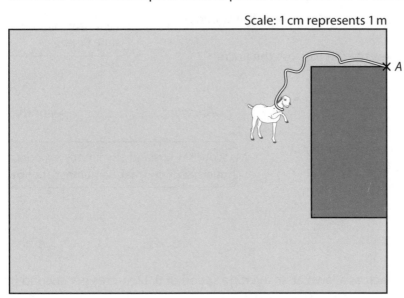

> Remember that the rope cannot go over or through the shed.

Draw the locus of all points that the goat can reach.

(3 marks)

 3 The diagram shows an L shape. Draw the locus of all points 2 cm from the L shape.

> Think carefully what happens at the ends of the L shape and at the right-angle.

(4 marks)

Speed

 1 A cyclist travels a distance of 64 miles in 4 hours.
What is the cyclist's average speed?

> **Guided**

$S = \dfrac{D}{T}$ ....

....

Answer .... mph   *(2 marks)*

> Draw the formula triangle for speed and use it to write out the formula you need.

 2 A plane travels for 7 hours at an average speed of 475 miles per hour. Work out the distance travelled by the plane.

> Use the formula triangle to write out the formula for distance.

> **Guided**

$D = S \times T =$ ....

....

Answer .... miles   *(2 marks)*

3 A girl runs 12 metres across a lawn at an average speed of 4 metres per second (m/s).
Work out the time it takes the girl to cross the lawn.

> First write out the formula for time taken.

....

Answer .... seconds   *(2 marks)*

4 A cyclist travels a distance of 78 miles in 3 hours and 15 minutes.
What is the cyclist's average speed?

> Write 3 hours and 15 minutes in hours as a mixed number or decimal. Remember, it's **not** 3.15.

....

Answer .... mph   *(3 marks)*

5 A baby crawls 3 metres across a lawn at an average speed of 0.02 metres per second (m/s).
Work out the time it takes the baby to cross the 3 metres of lawn in minutes.

....

Answer .... minutes   *(3 marks)*

 6 George drives 156 miles from his house to London.
He drives at an average speed of 60 miles per hour.
He leaves his house at 6:30 am.
Does he arrive in London before 9:00 am?
You **must** show your working.

> Work out the time taken for the journey, add that time onto 06:30 and then answer the question.

....

....

....

Answer ....   *(4 marks)*

Density

1 The diagram shows a block of wood in the shape of a cuboid.
The density of the wood is 0.6 grams/cm³.
Work out the mass of the block of wood.

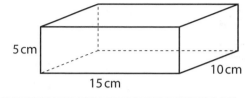

5 cm   10 cm   15 cm

> Start by working out the volume of the block. Draw the formula triangle for density before writing out the formula for mass.

Volume of block $= l \times w \times h$

$= 15 \times 10 \times 5 = \ldots\ldots\ldots\ldots$ cm³.

....

....

Answer .... grams   *(4 marks)*

2 The diagram shows a triangular prism.
The prism is made of glass.
The density of glass is
1.4 grams/cm³.
Work out the mass of
the prism.

5 cm   3 cm   20 cm

> The formula for the volume of a prism is on the formula sheet.

Volume of prism = area of cross-section × length

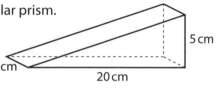

$= \ldots\ldots\ldots\ldots \times \ldots\ldots\ldots\ldots = \ldots\ldots\ldots\ldots$ cm³

....

....

Answer .... grams   *(4 marks)*

3 The diagram shows a block of wood in the shape of a cuboid.
The block of wood has a mass of 216 grams.
Work out the density of the block of wood.

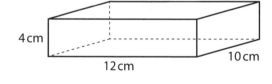

4 cm   12 cm   10 cm

....

....

Answer .... grams/cm³   *(4 marks)*

4 The diagram shows a pure gold pendant in the shape of a cuboid.
Pure gold has a density of 19.3 g/cm³ and a value of £31 per gram.
Work out the value of the pendant.

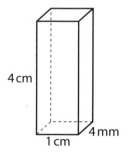

4 cm   1 cm   4 mm

....

....

....

Answer £.........................   *(4 marks)*

Converting units

D   **1**   Andrew is driving in Spain.
The speed limit is 100 km/h.

Guided   His speedometer shows he is travelling at 65 mph.
Is Andrew breaking the speed limit?
You **must** show your working.

> You need to remember the km to miles conversion.
> 5 miles = 8 km

Converting 100 km to miles = $100 \times \frac{5}{8}$ = ...

....   *(3 marks)*

D   **2**   Convert 108 km/h to m/s.

> Use multiplication to convert 108 km to metres and one hour to seconds.

....

....

....

Answer .... m/s   *(3 marks)*

D   **3**   Brenda spends 189 euros on fuel in France.
The fuel costs 1.20 euros per litre.

(a) How many litres of fuel does she buy?

....

Answer .... litres   *(2 marks)*

(b) 1 gallon = 4.5 litres. How many gallons of fuel does she buy?

....

Answer .... gallons   *(2 marks)*

She travels 1400 miles in France.

(c) Work out her fuel economy in miles per gallon.

....   *(3 marks)*

Answer .... miles per gallon   *(2 marks)*

D   **4**   The recommended distance between the viewer and the television is called the ideal viewing distance.
The ideal viewing distance is three times the screen size.
Garth buys a 50 inch television.
The sofa is 3.6 metres away from the television.
Is the sofa at the ideal viewing distance?
You **must** show your working.

> 1 inch = 2.5 cm

....

....

....

....

....   *(3 marks)*

Translations, reflections and rotations

1 Triangle *A* is shown on the grid.

> Guided

Translate triangle *A* by vector $\begin{pmatrix} -2 \\ 2 \end{pmatrix}$

> The top part of the vector is the *x* movement.

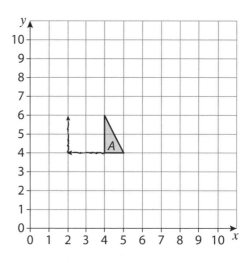

(2 marks)

2 The diagram shows two shapes *A* and *B*.

Describe fully the **single** transformation which takes shape *A* to shape *B*.

> Students have struggled with exam questions similar to this – **be prepared!**

EXAM ALERT

> You can use tracing paper to help answer this question.

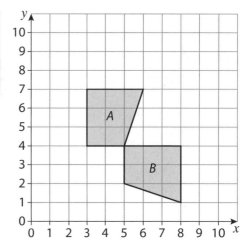

....

....

....

(3 marks)

3 The grid shows two shapes, A and B.

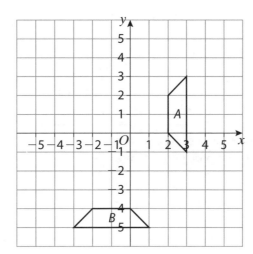

(a) Reflect shape *A* in the line $y = x$.   *(2 marks)*

(b) Describe fully the **single** transformation that takes shape *B* onto shape *A*.   *(3 marks)*

Enlargements

 1 The diagram shows the shape *ABCD* drawn on a grid.

Guided

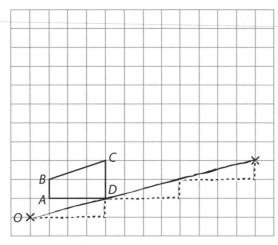

> Each point on your enlargement will be three times as far from *O* as the corresponding point on the original shape.

Enlarge *ABCD* by scale factor 3 using the point marked *O* as the centre of enlargement.

(2 marks)

A **2** Enlarge the shape by scale factor $-\frac{1}{2}$ with centre of enlargement $(-1, 0)$. *(3 marks)*

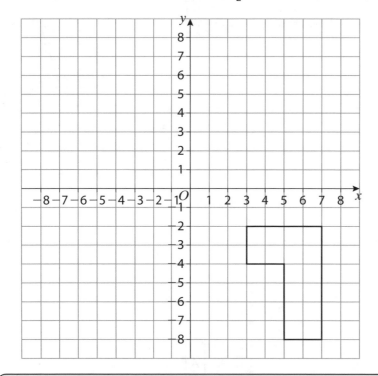

> Draw lines from each vertex through the centre of enlargement. The enlarged shape will be smaller than the original shape, on the other side of the centre of enlargement, and upside down.

Combining transformations

- To describe a translation you need to give a vector.
- To describe a reflection you write 'reflection' and give the equation of the line.
- To describe a rotation write 'rotation' and give the centre of rotation, and the angle and direction of turn.

1 The diagram shows triangle T.

(a) Rotate triangle T 90° clockwise about (1,3).
Label your new triangle A.   *(2 marks)*

(b) Rotate triangle A 90° anticlockwise about (3,2).
Label your new triangle B.   *(2 marks)*

(c) Describe fully the **single** transformation that takes
triangle B onto triangle T.

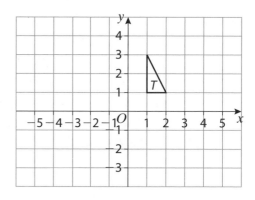

.... *(2 marks)*

2 The diagram shows triangle T.

(a) Reflect triangle T in the y-axis. Label your new
triangle A.   *(1 mark)*

(b) Rotate triangle A 90° anticlockwise about the origin.
Label your new triangle B.   *(2 marks)*

(c) Describe fully the **single** transformation that takes
triangle T onto triangle B.

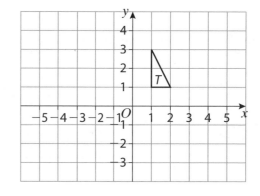

.... *(2 marks)*

3 The diagram shows triangle T.

(a) Reflect triangle T in the line $x = -1\frac{1}{2}$.
Label your new triangle A.   *(1 mark)*

(b) Translate triangle A by the vector $\begin{pmatrix} 2 \\ -1 \end{pmatrix}$.
Label your new triangle B.   *(2 marks)*

(c) Reflect triangle B in the line $y = x$.
Label your new triangle C.   *(2 marks)*

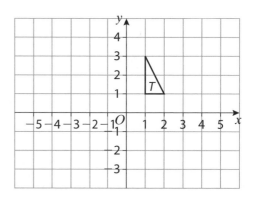

(d) Describe fully the **single** transformation that
takes triangle C onto triangle T.

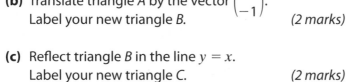

.... *(3 marks)*

Similar shapes 2

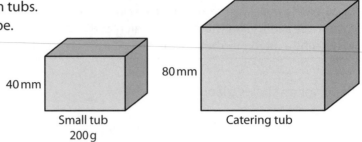

A **1** Low fat margarine is sold in tubs.
The tubs are similar in shape.

> **Guided**

Small tub
200 g

Catering tub

Not drawn
accurately

40 mm

80 mm

(a) The dimensions of the catering tub are double those of the small tub.
The small tub holds 200 grams of margarine.
How much does the catering tub hold?

Linear scale factor = 2

Volume scale factor = 2^3 = ...

New capacity = 200 × .... = ...

Answer .... g   *(2 marks)*

A* **(b)** A large tub holds 800 grams of margarine.

> **Guided**

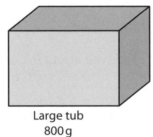

40 mm

Small tub
200 g

Large tub
800 g

Not drawn
accurately

Students have
struggled with exam
questions
similar to this –
be prepared!

The height of the 200 gram size tub is 40 mm.
Work out the height of the large tub.

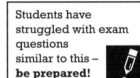

The **volume** multiplier from
200 g to 800 g is 4. Use this to
work out the height multiplier.
The height multiplier won't be
a whole number.

Volume multiplier (k^3) = $\dfrac{800}{200}$ = 4

Height multiplier (k)   = ...

....

....

Answer .... mm   *(3 marks)*

A* **2** The diagram shows two similar solid cones. The area of
the base of each cone is given.
Work out the value of x.

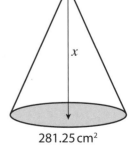

....

....

....

....

5 cm

x

125 cm²

281.25 cm²

Answer .... cm   *(3 marks)*

Vectors

1 In the diagram, triangle PQR has a point M such that $QM = \frac{3}{4}QR$.
$\overrightarrow{PQ} = \mathbf{a}$ and $\overrightarrow{PR} = \mathbf{b}$.

Guided

Work out $\overrightarrow{PM}$ in terms of $\mathbf{a}$ and $\mathbf{b}$.
Give your answer in its simplest form.

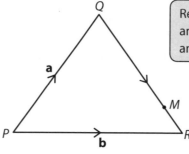

Remember to simplify the answer by expanding brackets and collecting like terms.

$$\overrightarrow{QR} = -\mathbf{a} + \mathbf{b} = \mathbf{b} - \mathbf{a}$$

$$\overrightarrow{QM} = \frac{3}{4}\overrightarrow{QR} = \frac{3}{4}(\mathbf{b} - \mathbf{a})$$

$$\overrightarrow{PM} = \overrightarrow{PQ} + \overrightarrow{QM}$$

$$= ....$$

Answer .... *(3 marks)*

2 $ABCD$ is a trapezium as shown.
AB is parallel to DC.
$DC = 2AB$.

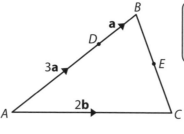

$\overrightarrow{DC}$ is in the same direction, but twice as long as $\overrightarrow{AB}$.

(a) Write down vector DC in terms of $\mathbf{s}$ and $\mathbf{t}$.

Answer .... *(1 mark)*

(b) Work out vector BC in terms of $\mathbf{s}$ and $\mathbf{t}$.
Give your answer in its simplest form.

To get from B to C you have to go from B to A to D to C. Remember that the direction of the arrow is important. If $\overrightarrow{AB} = \mathbf{s}$, then $\overrightarrow{BA} = -\mathbf{s}$.

....

....

....

Answer .... *(2 marks)*

3 In the diagram, $\overrightarrow{AC} = 2\mathbf{b}$, $\overrightarrow{AD} = 3\mathbf{a}$,
$\overrightarrow{DB} = \mathbf{a}$ and E is the midpoint of BC.
Find $\overrightarrow{DE}$ in terms of $\mathbf{a}$ and $\mathbf{b}$.

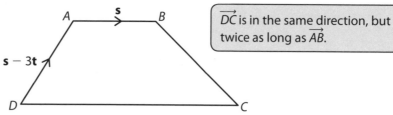

First work out the vector $\overrightarrow{BE}$ (which is half of $\overrightarrow{BC}$), and add it onto $\overrightarrow{DB}$.

....

....

....

....

Answer .... *(3 marks)*

Solving vector problems

1 The diagram shows a parallelogram *ABCD*.
$\overrightarrow{AB} = \mathbf{a}$ and $\overrightarrow{AD} = \mathbf{b}$.

Guided

N is a point on *BC* such that $BN:NC = 2:1$.

M is the midpoint of *CD*.

Work out, in terms of **a** and **b**, expressions for the following vectors, simplifying your answers where possible.

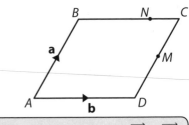

(a) $\overrightarrow{AM} = \overrightarrow{AD} + \overrightarrow{DM}$

= b + ...

> As this is a parallelogram, $\overrightarrow{AB} = \overrightarrow{DC}$ and *M* is half way along $\overrightarrow{DC}$.

Answer .... *(1 mark)*

(b) $\overrightarrow{AN} = \overrightarrow{AB} + \overrightarrow{BN}$

= a + ...

> As this is a parallelogram, $\overrightarrow{AD} = \overrightarrow{BC}$. Use the ratio given to work out what **fraction** *N* is along *BC*.

Answer .... *(1 mark)*

(c) $\overrightarrow{NM}$ ....

....

....

....

Answer .... *(1 mark)*

2 The diagram shows triangle *ABC*:

D is the midpoint of *AB*.

Points *E* and *F* are on *BC* such that $BE = EF = FC$.

$\overrightarrow{AD} = \mathbf{a}$ and $\overrightarrow{AC} = \mathbf{b}$.

(a) Find, in terms of **a** and **b**, expressions for the following vectors:

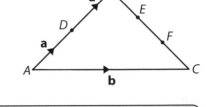

(i) $\overrightarrow{BC}$ ....

> To get from *B* to *C* you could travel $B \rightarrow A \rightarrow C$.

Answer .... *(1 mark)*

(ii) $\overrightarrow{DE}$ ....

....

> You have the vector for $\overrightarrow{BC}$, and you know that *BE* is $\frac{1}{3}$ of the way along *BC*.

....

Answer .... *(2 marks)*

(iii) $\overrightarrow{AF}$ ....

....

Answer .... *(2 marks)*

(b) What type of quadrilateral is *ADEF*? Give a reason for your answer.

....

.... *(2 marks)*

Problem-solving practice 1

1 The diagram shows parts of two regular polygons A and B.

A has 8 sides and exterior angle $3x$.
B has exterior angle $2x$.
Work out the number of sides of regular polygon B.

Not drawn accurately

> The sum of the exterior angles of any polygon is 360°. Use this fact to work out x.

Exterior angle of A = $\dfrac{360°}{8}$ = ...

$3x$ = ...

x = ...

....

....

....

Answer ....   *(4 marks)*

2 The diagram shows a semi-circle, radius r, and a rectangle.
The perimeters are equal.
Work out the value of r.
Give your answer correct to 2 decimal places.

Not drawn accurately

r cm

r cm   4 cm

....

> Use the information given in the question to write an equation. Solve your equation to find r.

....

....

....

Answer r =........................ cm   *(2 marks)*

3 Two banks calculate the yearly interest they pay customers.

Bank A: 3% of the total that you invest	**Bank B:** 1% of the first £300 that you invest
For example: Invest £1000	5% of amounts over £300 that you invest
Interest = 3% of £1000	For example: Invest £1000
	Interest = 1% of £300 + 5% of £700

Let x be the amount of money that will earn the same yearly interest at the two banks.
Use the information to set up and solve an equation to find x.
Do **not** use trial and improvement.

$0.03 \times x = 0.01 \times 300 +$ ....

....

....

....

Answer £........................   *(3 marks)*

Problem-solving practice 2

4 *ABCD* is a cyclic quadrilateral as shown.
Prefix: **Guided**

Prove that *AD* is parallel to *BC*.

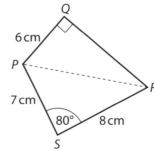

$9x - 40° + 3x - 20° = 180°$

....

....

....

....

....

....

....

....

....

> Opposite angles of a cyclic quadrilateral add up to 180°. Start by using this information to write an equation.

(4 marks)

5 Here is a quadrilateral, *PQRS.*°
PQ = 6 cm, *PS* = 7 cm and *RS* = 8 cm.
Angle *PSR* = 80° and angle *PQR* = 90°.
Work out the area of the quadrilateral *PQRS*.

> You will need to work out *PR* first, using the cosine rule. Then work out *QR* using Pythagoras' theorem.

....

....

....

....

....

....

....

....

....

....

....

....

Answer ....   *(6 marks)*

Unit 1 Practice Exam Paper

This practice exam paper has been written to help you practise what you have learned and may not be representative of a real exam paper.

Time allowed: 1 hour
You may use a calculator.

1 Jordan measures the distance she can throw balls of different masses.
The table shows her results.

Distance (nearest metre)	2	8	6	10	13	12
Mass of ball (grams)	2700	1300	1900	900	500	600

(a) On the grid, draw a scatter diagram to show this information.

(2 marks)

(b) Write down the type of correlation shown by your scatter diagram.

Answer .... *(1 mark)*

(c) Draw a line of best fit on your scatter diagram.
(1 mark)

(d) Jordan has one more ball to throw.
The ball has a mass of 1500 g.
Use your line of best fit to estimate how far she could throw this ball.

Answer .... m *(1 mark)*

2 Nick added four identical mixed numbers. He got an answer of $11\frac{1}{3}$.
What were the mixed numbers that Nick added?

....

Answer .... *(2 marks)*

3* The cost of a computer is £360 + 20% VAT.
Roz decides to pay on credit.
She pays an initial deposit of 15% and then 12 monthly payments of £35.
How much more does Roz pay by buying on credit?
You **must** show your working.

....
....
....

Answer £......................... *(4 marks)*

4 Basir runs a 'hoopla' game at a charity fete.
The probability of someone winning the £50 prize is $\frac{1}{120}$.
The rest of the time it pays out nothing.
360 people play the game.

(a) How much prize money should Basir expect to pay out in total?

....
....
....

Answer £......................... *(3 marks)*

(b) It costs 50p to play the hoopla game once.
Is Basir likely to make a profit?
You **must** show your working.

....

....

Answer .... *(2 marks)*

5 Amy follows this recipe for maxi cup cakes.

> **Maxi Cup Cakes (makes 9)**
> 120 g butter
> 120 g caster sugar
> 120 g self-raising flour
> 2 free range eggs

She uses 200 g of butter.
How many maxi cup cakes can she make?

....

Answer .... *(2 marks)*

6 The frequency table gives information of the time taken by 50 students to complete a puzzle.

Time taken, t (seconds)	Frequency	
$30 < t \leqslant 40$	8	
$40 < t \leqslant 50$	18	
$50 < t \leqslant 60$	13	
$60 < t \leqslant 70$	7	
$70 < t \leqslant 80$	4	

(a) Which class interval contains the median?

Answer .... *(1 mark)*

(b) Work out an estimate for the mean time taken to complete the puzzle.

....

....

....

....

....

Answer .... seconds *(4 marks)*

(c) Explain why you can't work out an exact mean time taken to complete the puzzle.

....

....

.... *(1 mark)*

7* Pauline buys a van for £2250. She spends £350 on new tyres and then sells the van for £3250.
Pauline says that she makes 25% profit. Is Pauline correct?
You **must** show your working.

....

....

....

....

Answer .... *(3 marks)*

8 The distance a cricket ball was thrown by 50 students was measured.
The frequency table shows the results.

Distance, d (metres)	Frequency
$14 < d \leqslant 18$	3
$18 < d \leqslant 22$	10
$22 < d \leqslant 26$	13
$26 < d \leqslant 30$	21
$30 < d \leqslant 34$	3

(a) Draw a cumulative frequency graph for this data.

(4 marks)

(b) Use your graph to estimate:

(i) the median

....

Answer .... m *(1 mark)*

(ii) the interquartile range

....

....

Answer .... m *(2 marks)*

The distance a rounders ball was thrown
by 50 students was also measured.
The box plot shows the distances thrown
for the rounders ball.
The minimum distance thrown for the
cricket ball was 15 metres.
The maximum distance thrown for the
cricket ball was 33 metres.

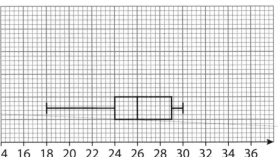

14 16 18 20 22 24 26 28 30 32 34 36

(c) Draw a box plot for the cricket ball above the box plot for the rounders ball. *(2 marks)*

(d) Another student is asked to throw either a cricket ball or a rounders ball.
Which type of ball should they choose if they want to throw the ball as far as possible?
Explain your answer.

....

....

....

.... *(2 marks)*

9 The table shows the ages of the members of a scout group.

Age (years)	10	11	12	13	14
Number of members	11	43	37	31	28

The scout leader wants a stratified sample of 30 members.
How many members should be chosen from each age group?

....

....

....

Answer .... *(3 marks)*

10 Alan, Barry and Ceri are to take a penalty kick.
The probability that Alan will score is 0.9
The probability that Barry will score is 0.3
The probability that Ceri will score is 0.6
Work out the probability that:

(a) all of them score?

....

....

Answer .... *(2 marks)*

(b) exactly two of them score?

....

....

....

Answer .... *(3 marks)*

11 This histogram represents the lengths of time golfers spent on a golf course one day.

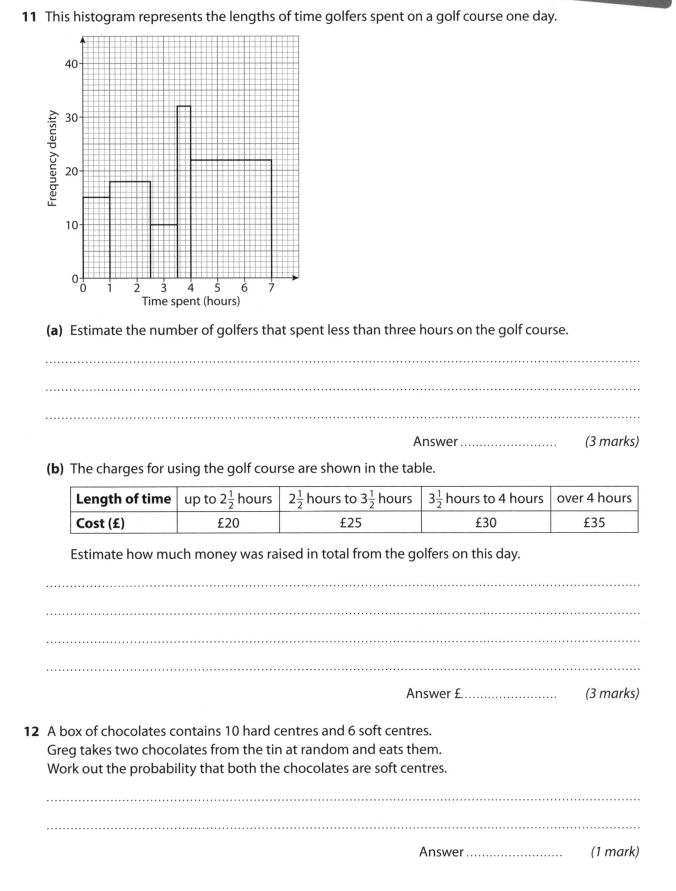

(a) Estimate the number of golfers that spent less than three hours on the golf course.

....

....

....

Answer .... *(3 marks)*

(b) The charges for using the golf course are shown in the table.

Length of time	up to $2\frac{1}{2}$ hours	$2\frac{1}{2}$ hours to $3\frac{1}{2}$ hours	$3\frac{1}{2}$ hours to 4 hours	over 4 hours
Cost (£)	£20	£25	£30	£35

Estimate how much money was raised in total from the golfers on this day.

....

....

....

....

Answer £......................... *(3 marks)*

12 A box of chocolates contains 10 hard centres and 6 soft centres.
Greg takes two chocolates from the tin at random and eats them.
Work out the probability that both the chocolates are soft centres.

....

....

Answer .... *(1 mark)*

Unit 2 Practice Exam Paper

This practice exam paper has been written to help you practise what you have learned and may not be representative of a real exam paper.

Time allowed: 1 hour 15 minutes
You **must not** use a calculator.

1 Use approximations to estimate the value of $\dfrac{4.15 \times 18.8}{9.11 - 4.7}$

....

....

....

Answer .... *(3 marks)*

2 Here is a number machine. **Input** ⟶ Add 3 ⟶ Multiply by 2 ⟶ **Output**

(a) Work out the output when the input is 5.

....

Answer .... *(1 mark)*

(b) When the input is n what is the output?

....

Answer .... *(2 marks)*

3* A gas meter measures the number of units of gas a person uses.
In 2011 a unit of gas cost 5 pence.
In 2011 John used 10 000 units of gas.
In 2012 a unit of gas cost 20% more than it did in 2011.
In 2012 John used 20% fewer units of gas than he did in 2011.
Will John's gas bill be the same in 2012 as it was in 2011?
You **must** show your working.

....

....

....

....

....

Answer .... *(4 marks)*

4 Sam answers a questionnaire about weight.
He ticks this box to show what his weight in kilograms lies between.

✓ $70 \leqslant w < 75$

(a) Write down all the whole-number kilogram weights that Sam could weigh.

Answer .... *(1 mark)*

(b) Show the inequality $70 \leqslant w < 75$ on this number line.

68   69   70   71   72   73   74   75   76   77   78

(1 mark)

5* Kevin is laying tiles on his kitchen floor.
The tiles are going to cover a rectangular area measuring 6 m by 3.6 m.
Tiles come in packs that cost £20 per pack. Each pack covers an area of 1 m².
Kevin is not good at DIY so decides to buy more packs than he should need.
He buys an extra 10% of the tiling area.
How much does Kevin pay for his tiles?

....

....

....

....

....

Answer £......................... *(3 marks)*

6 An electricity supplier uses this formula to work out the total cost of the electricity a customer uses.

$C = 45 + 0.09k$

where C is the total cost of the electricity in pounds and k is the number of kilowatt-hours of electricity used.

(a) Jason uses 3000 kilowatt-hours of electricity.
What is the total cost of the electricity he uses?

....

....

Answer £......................... *(2 marks)*

(b) The total cost of the electricity Lisa uses is £270.
How many kilowatt-hours of electricity does she use?

....

....

Answer .... kWh *(2 marks)*

7 **(a)** Solve the equation $8x - 12 = 9 + 2x$

....

....

....

Answer $x = $ .... *(3 marks)*

(b) Expand and simplify $3(2y + 1) - 2(y - 5)$

....

....

....

Answer .... *(3 marks)*

(c) Solve these simultaneous equations.

$3x + 2y = 13$

$5x - 3y = 9$

....

....

....

....

....

$x = $ .... $y = $ .... *(4 marks)*

8 The nth term of a sequence is $4n + 2$

(a) Show that all the terms in the sequence are even.

....

....

.... *(2 marks)*

(b) How many terms of the sequence lie between 150 and 250?
You **must** show your working.

....

....

....

....

Answer .... *(3 marks)*

9 By drawing suitable straight lines on the grid below, shade the region which satisfies these inequalities:

$y \leqslant 1$ $\quad x \geqslant -2$ $\quad y \leqslant 3x - 1$

(4 marks)

10 Harry plants a handful of seeds.

70% of the seeds are tomato seeds and 30% of the seeds are cucumber seeds.

90% of the tomato seeds germinate and 80% of the cucumber seeds germinate.

What percentage of the handful of seeds germinate?

....

....

....

....

Answer .... %   *(3 marks)*

11 P is the point $(1, -2)$ and Q is the point $(10, 16)$.

A straight line parallel to PQ passes through the point $(6, 5)$.

Work out the equation of this line.

You **must** show your working.

....

....

....

....

Answer ....   *(3 marks)*

12 Simplify fully $\dfrac{2x^2 + x - 15}{4x^2 - 25}$

....

....

....

....

....

....

Answer ....   *(5 marks)*

13 (a) Simplify fully $\dfrac{(x^4)^3}{x^8}$

....

....

....

Answer ....   *(2 marks)*

(b) Simplify $(2a^3b^2)^4$

....

....

....

Answer ....   *(2 marks)*

(c) To find the value of $125^{-\frac{1}{3}}$ Kim wrote:
$125^{\frac{1}{3}} = 5$, so the answer is -5.
Is Kim correct? Explain your answer.

....

....

.... *(2 marks)*

14 (a) Write the expression $x^2 - 8x + 8$ in the form $(x + a)^2 + b$.

....

....

....

Answer .... *(3 marks)*

(b) Hence, or otherwise, solve $x^2 - 8x + 8 = 0$.
Give your answers in surd form.

....

....

Answer .... *(2 marks)*

15 Write the expression $(\sqrt{80} - 1)(\sqrt{20} + 3)$ in the form $a + b\sqrt{5}$, where a and b are integers.

....

....

....

....

Answer .... *(4 marks)*

Unit 3 Practice Exam Paper

This practice exam paper has been written to help you practise what you have learned and may not be representative of a real exam paper.

Time allowed: 1 hour 30 minutes
You may use a calculator.

1* Carrick buys and sells cups. He buys a box of 80 cups for £55.
He sells all of the cups for £1.25 each.
How much profit does Carrick make on this box of cups?

....

....

Answer .... *(2 marks)*

2 The diagram shows a quadrilateral.
Work out the value of a.

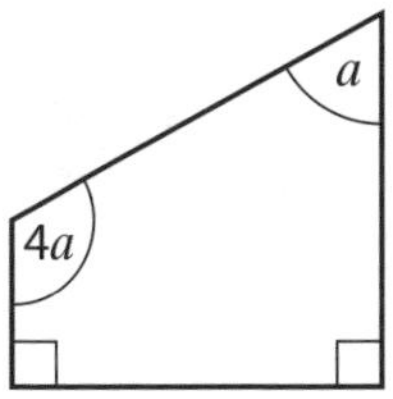

a
4a
Not drawn accurately

....

$a =$ .... *(2 marks)*

3 *AB* and *CD* are parallel lines.
Work out the sizes of angles x and y.
You must give reasons for your answers.

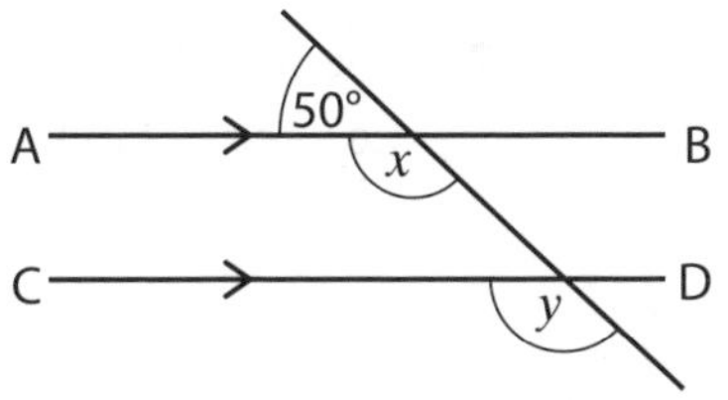

Not drawn accurately
50°
A
B
x
C
D
y

....

....

....

.... *(3 marks)*

4 The diagram shows a shape made from trapezium *ABEF* and rectangle *BCDE*.

Calculate the area of the whole shape.

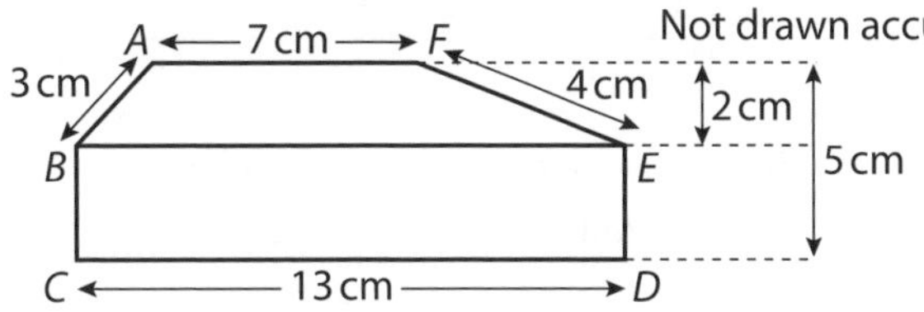

Not drawn accurately
A
7 cm
F
3 cm
4 cm
2 cm
B
E
5 cm
C
13 cm
D

....

....

....

....

Answer .... cm^2 *(5 marks)*

5 Dianne has £x. Ethan has £5 more than Dianne. Fatima has twice as much as Dianne.
T is the total of all of their money.

(a) Write a formula for *T* in terms of x.
Give your answer in its simplest form.

....

....

Answer .... *(2 marks)*

(b) The total amount of money all three people have is £57. How much does Ethan have?

....

....

Answer £......................... *(3 marks)*

6 An ant leaves home and walks on a bearing of 055° for 4 m to a point *A*.
At *A* the ant changes direction and walks 5 m on a bearing of 210° to a point *B*.

N

(a) Using a scale of 1 cm = 1 m, draw a diagram to show the ant's journey.

•
Home

(2 marks)

(b) Measure and write down the distance and the bearing the ant must walk on to return directly from point *B* to home.

....

Answer .... *(2 marks)*

7 You are given that 5 miles = 8 kilometres.
Convert 12 miles into kilometres.

....

....

Answer .... km *(3 marks)*

8 **(a)** Complete the table of values for $y = x^2 + 1$ for values of *x* from −3 to +3.

x	−3	−2	−1	0	1	2	3
y	10		2	1		5	

(2 marks)

(b) Draw the graph of $y = x^2 + 3$ for values of *x* from −3 to +3.

(2 marks)

(c) Use your graph to estimate the value of *y* when $x = -2.5$.

Answer *y* = ... *(1 mark)*

9 **(a)** Triangle B is an enlargement of triangle A.

    (i) Write down the scale factor of the enlargement.

      Answer .... *(1 mark)*

    (ii) Write down the coordinates of the centre of the enlargement.

      Answer .... *(1 mark)*

(b) Enlarge triangle C by scale factor $\frac{1}{3}$, using (3, 7) as the centre of enlargement.

(2 marks)

10 Using only compasses and a ruler, construct the perpendicular from the point *P* to the line. You **must** show your construction lines.

• P

_____

(2 marks)

11 The area of this rectangle is 50 cm².
Use trial and improvement to find the value of x correct to 1 d.p.

$(x^2 - 1)$ cm

x cm

....

....

....

....

Answer $x =$ .... *(4 marks)*

12 Solve the equation $(x + 6)(x - 1) = (x + 5)(x + 3)$

....

....

....

Answer $x =$ .... *(3 marks)*

13 In the diagram $BC = 10$ cm, $AD = 30$ cm and angle $DAB = 25°$.

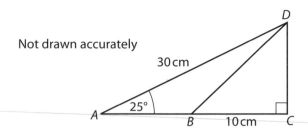

Not drawn accurately

Calculate:

(a) the length of CD

....

Answer .... cm   *(2 marks)*

(b) the size of angle CBD

....

Answer .... degrees   *(2 marks)*

14 A man is 180 cm tall. He casts a shadow 3 m long.
At the same time a tree casts a shadow 55 m long.
Show that the height of the tree is 33 m.
You **must** show your working.

....

....

....

....   *(3 marks)*

15 A, B, C, D and E are points on the circumference of a circle centre O.
BE is a diameter of the circle.

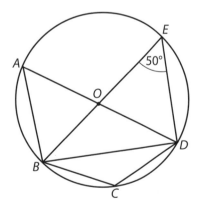

(a) Write down the size of angle BAD.

Answer .... degrees   *(1 mark)*

(b) Work out the size of angle DBE.

....

Answer .... degrees   *(1 mark)*

(c) Work out the size of angle BCD.

....

Answer .... degrees   *(1 mark)*

16 Zac and Adam start with the same negative number.
Zac subtracts 3 from the number and then multiplies the result by 2.
Adam squares the number and then subtracts 41 from the result.
They both finish with the same answer.
Work out the negative number they started with.
You **must** show your working.

....

....

....

....

....

Answer .... *(4 marks)*

17 *ABC* is a tangent to the circle at *B*.
O is the centre of the circle.
Angle *BOD* = $2x$
Show that angle *CBD* = x

Not drawn
accurately

....

....

....

....

....

Answer .... *(3 marks)*

18 *M* is inversely proportional to the square of x.
When $M = 4$, $x = 6$.
Work out the value of *M* when $x = 3$

....

....

....

....

Answer .... *(4 marks)*

19 (a) Vector $\mathbf{a} = \begin{pmatrix} 2 \\ -4 \end{pmatrix}$, $\mathbf{b} = \begin{pmatrix} 2 \\ -1 \end{pmatrix}$ and $\mathbf{c} = \mathbf{a} - 2\mathbf{b}$

On the grid below draw a diagram to illustrate the vector **c**.

(2 marks)

(b) *ABCDEF* is a regular hexagon.

$\overrightarrow{AB} = \mathbf{a}$ and $\overrightarrow{BC} = \mathbf{b}$

Work out the following vectors in terms of **a** and **b**.
Simplify your answers where possible.

(i) $\overrightarrow{ED}$

....

Answer .... *(1 mark)*

(ii) $\overrightarrow{EF}$

....

Answer .... *(1 mark)*

(iii) $\overrightarrow{CD}$

....

Answer .... *(1 mark)*

(iv) $\overrightarrow{CE}$

....

Answer .... *(1 mark)*

20* A sphere has surface area 9π cm^2.

(a) Show that the radius of the sphere is 1.5 cm.

....

....

....

.... *(3 marks)*

(b) Work out the volume of the sphere.
Give your answer in terms of π.

....

....

Answer .... cm^3 *(2 marks)*

(c) 12 of these spheres are melted down and recast into a cylinder of height 8 cm.
Work out the radius of the cylinder.
Give your answer correct to 1 decimal place.

....

....

....

Answer .... cm *(3 marks)*

21 (a) On the axes below, sketch the graph of $y = \sin x$ for values of x in the range $0° \leqslant x \leqslant 360°$

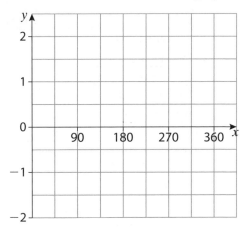

(b) On the axes below, sketch the graph of $y = -\sin x$

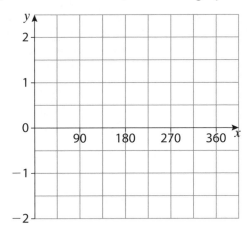

(1 mark)

(c) On the axes below, sketch the graph of $y = 1 + \sin x$

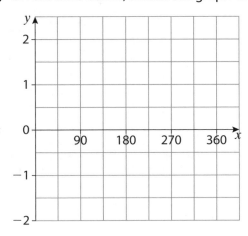

(1 mark)

Answers

Unit 1

1. Calculator skills 1
1. (a) 4.26315…   (b) 4.3
2. (a) 2.56652…   (b) 2.57
3. $\frac{25}{8}$ $\frac{22}{7}$ $\frac{54}{17}$
4. 0.345   35% $\frac{9}{25}$ $\frac{3}{8}$
5. £9.75
6. 38.4%

2. Percentage change 1
1. £13 442.10
2. £712
3. Yes (£95.58)
4. 24%
5. 85.5 cm, 104.5 cm

3. Reverse percentages and compound interest
1. £5624.32
2. £115
3. 4 hours
4. Less, 0.64%

4. Ratio
1. 2 : 5
2. 1 : 2 : 4
3. 1 : 2.25
4. *Example answer:* 4 : 10 or 6 : 15
5. £8 to £40
6. 42
7. 190 g

5. Standard form
1. 3.5×10^7
2. 601 000
3. 8.09×10^5
4. 4.065×10^{13} km
5. (a) 1.555×10^{11} (b) 1.089×10^{-15} (c) 1.21×10^{-9}
6. 7.314×10^{-23} kg

6. Upper and lower bounds
1. (a) 1565.5 ml   (b) 1535.5 ml
2. 66.0 people per square km
3. 293.25 m

7. The Data Handling Cycle
1. *Example answer:* Collect data by looking for prices online at internet sites and then looking for prices at my local high street stores for the same type of iPod. Calculate the mean and median prices for the iPod in internet sites and local high street stores. Compare the averages for internet sites and local high street stores and write a conclusion.
2. *Example answer:* The hypothesis is: the number of students who choose a healthy option of Fridays is 20% more than other days. Collect data by giving a questionnaire to a random sample of 10% of children in my school. The questionnaire will ask what days they eat in the canteen and what days they choose the 'healthy option'. Calculate the percentage of students who have the healthy option each day. Compare the percentages and write a conclusion.
3. *Example answer:* Collect data by giving a questionnaire to a random sample of 10% of children in my school. The questionnaire will ask students to write down the length of time they spend doing homework each day for one week. Calculate the mean and median time spent for boys and for girls. Compare the averages and write a conclusion.

8. Collecting data
1.

Watch how people react to a loud noise	Data Logging
Obtain opinions on a new perfume	Observation
Collect data on reaction times of students	Experiment
Record number of people through a turnstile	Questionnaire

2. Discrete, specific values only.
3. Continuous, any value possible.

4.

Data	Type of data
Number of trees in orchard	quantitative, discrete
Species of apple tree	qualitative
Diameter of tree trunk	quantitative, continuous
Number of apples per tree	quantitative, discrete
Weight of apples per tree	quantitative, continuous
Colour of apple	qualitative

5. Collecting data this way is collecting someone else's primary data. This data collection method means it is secondary data.

9. Surveys
1. *Example answer:* Do you study GCSE Geography?
   Yes ☐   No ☐
   Can you point to where Newcastle is on a blank map of the UK?
   Yes ☐   No ☐
2. (a) *Example answer:* Does not ask about other days of the week.
   (b) *Example answer:*   Zero ☐   less than 1 hour ☐
   1 hour to 2 hours 59 mins ☐   more than 3 hours ☐
3. *Example answer:* From the time you leave your house in the morning, on average, how many minutes does it take you to get to school.
   Example answer: 0–10 ☐   11–20 ☐
   21–30 ☐   31–40 ☐
   more than 40 ☐

10. Two-way tables
1. (a) 17 Female, Sports Studies; 42 Total, Sports Studies
   (b) 10
   (c) 42%
2. (a)

	Sports	Library	Chatting	Total
Male	8	12	13	33
Female	21	16	5	42
Total	29	28	18	75

   (b) 5   (c) $\frac{12}{75}$ or $\frac{4}{25}$
3.

	Male	Female	Total
Thought test was fair	12	18	30
Thought test was not fair	36	24	60
Total	48	42	90

11. Sampling
1. Runners may be biased towards a running track.
2. *Example answers:* Only choose students from your school. Every student must have an equal chance of being selected.
3. Where the sample groups are in the same proportion as the population.
4. 74
5.

Male	35
Female	15

12. Mean, median and mode
1. $10 \leqslant l < 12$
2. $14 \leqslant l < 16$
3. (a) 5 5 7 **8** **9** (any two different numbers > 7)
   (b) 5 5 7 **9** **90** (any two different numbers > 7)
4. 9
5. Because 28 is so much larger than the other numbers it will distort the mean.
6. *Example answer:* 0 1 1 1 1 2

13. Frequency table averages
1. 2.5
2. £104.46

14. Interquartile range
1. 8 cm
2. 11 s
3. Median = 38 m, IQR = 8 m

15. Scatter graphs

1 (a), (b)

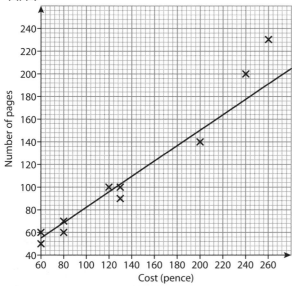

(c) Positive correlation. **(d)** More pages = more cost.
(e) Read off from line of best fit from graph above, 214 pence.

2 (a), (b)

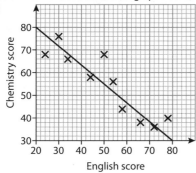

16. Frequency polygons

1 (a)

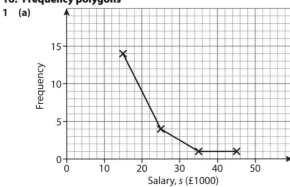

(b) £19 500

17. Histograms

1 (a)
(b) 49.1%

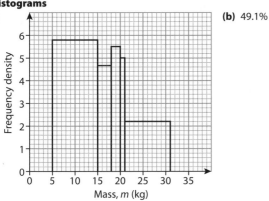

2 (a)

Speed (mph)	Frequency	Frequency density
$15 \leqslant s < 20$	90	18
$20 \leqslant s < 25$	130	26
$25 \leqslant s < 29$	120	30
$29 \leqslant s < 36$	105	15
$36 \leqslant s < 40$	40	10

Histogram of speed of cars

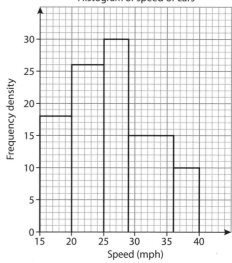

18. Probability 1

1 (a) 0.9   **(b)** 0.5   **(c)** 0.4

2 (a)

	1	2	3	4
H				
T				

(b) $\frac{1}{8}$ or 0.125

3 (a) 0.5   **(b)** 0.06

4 0.027

19. Probability 2

1 30

2 (a) $\frac{7}{30}$ or $0.23\dot{3}$ **(b)** $\frac{10}{30}$ or $0.3\dot{3}$

3 (a) $\frac{1}{9}$

(b) Might not be very accurate as 45 is a small sample.

4 160

20. Tree diagrams

1 (a) First counter   Second counter

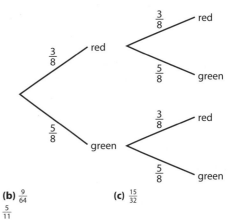

(b) $\frac{9}{64}$ **(c)** $\frac{15}{32}$

2 $\frac{5}{11}$

3 0.92

21. Cumulative frequency

1 (a)

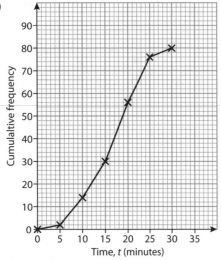

(b) 17 minutes   **(c)** 9 minutes
(d) Yes, $\frac{24}{80} \times 100 = 30\%$
(e) 22 minutes

22. Box plots

1 (a)

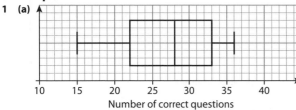

(b)

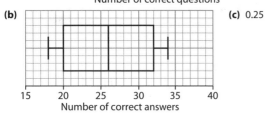

     (c) 0.25

23. Comparing data

1 (a) 60 seconds
(b) *Example answer:* On average the Saturday morning customers were on hold for a shorter time that the Tuesday morning customers (lower median: Sat = 40 secs, Tue = 45 secs). Tuesday morning customers holding time was more closely grouped than Saturday morning customers (smaller IQR: Tue = 60 secs, Sat = 85 secs).

2

	Median (grams)	IQR (grams)
Farm A	140	50
Farm B	100	60

Example answer: On average the oranges from Farm A were heavier than from oranges from Farm B (larger median: A = 140 g; B = 100g). The oranges from Farm A had a more consistent mass than from oranges from Farm B (smaller IQR: A = 50 g; B = 60 g).

24. Problem-solving practice 1

1 £559.50
2 $a = 17, b = 26$
3 Offer 1 = £8785; Offer 2 = £8712; Offer 3 = £8750; so Offer 2 is cheapest.
4 20, 40

25. Problem-solving practice 2

5 (a) 1.2×10^{12} litres   **(b)** 1.12×10^{12} litres
6 0.3125
7 $\frac{1}{24}$

Unit 2

26. Factors and primes

1 $2^4 \times 7$
2 (a) $2^5 \times 5$ **(b)** 8   **(c)** 480
3 Any two of 8, 24, 40, 56, …

27. Fractions and decimals

1 0.77   79% $\frac{17}{20}$
2 $\frac{3}{40}$ 9%   0.11
3 $\frac{7}{18}$
4 $1\frac{5}{21}$
5 $7\frac{5}{8}$
6 $2\frac{19}{24}$
7 $2\frac{11}{30}$

28. Decimals and estimation

1 (a) 236   **(b)** 96.76
2 (a) 78   **(b)** 3.25   **(c)** 812.5
3 5
4 1200
5 250
6 20

29. Recurring decimals

1 0.35
2 0.0875 or the denominator has prime factors of 2 and/or 5 *only* ($2^4 \times 5$), so terminating.
3 $0.1\dot{2}$ or the denominator has prime factors *other* than 2 and/or 5 ($3^2 \times 11$), so recurring.
4 $0.58\dot{3}$ or the denominator has prime factors *other* than 2 and/or 5 ($2^2 \times 3$), so recurring.
5 0.571428
6 (a) Long division shown, at least 3 decimal places calculated.
     (b) $\frac{2}{3} + \frac{1}{10} = \frac{23}{30}$

30. Percentage change 2

1 **A** = £180, **B** = £180.50, **A** is cheaper.
2 **A**, £2 extra.
3 Yes (cost = £230)

31. Ratio problems

1 Yes (£80 spare).
2 210 g
3 225 kg
4 75 ml

32. Indices 1

1 (a) -1 **(b)** 1   **(c)** -1 **(d)** 1   **(e)** -1
2 No, $4^{12} \div 4^2 = 4^{10}$
3 (a) 2^6 **(b)** 5^2 **(c)** 7^6
4 (a) 2^4 **(b)** 5^3
5 (a) 3^{-2} **(b)** 9^{-4} **(c)** 2^{12}

33. Indices 2

1 $\frac{1}{27}$
2 $\frac{3}{2}$
3 1.5×10^{-1} $(0.15 > 0.125)$
4 $\frac{25}{4}$
5 (a) 9   **(b)** 1000   **(c)** 4
     (d) $\frac{1}{4}$
6 125
7 48

34. Surds

1 (a) $\sqrt{21}$ **(b)** $\sqrt{7}$ **(c)** $\sqrt{3}$ **(d)** $5\sqrt{2}$ **(e)** $3\sqrt{3}$
2 $7\sqrt{2}$
3 $11 - 6\sqrt{2}$ so, $a = 11, b = -6$
4 $2\sqrt{15}$
5 $3\sqrt{2}$

35. Algebraic expressions

1 (a) y^9 **(b)** x^4 **(c)** p^6
     (d) t^{10} **(e)** m^{10}
2 (a) y^{-9} **(b)** x^{-2} **(c)** p^{-10}
     (d) t^{20}
3 (a) $5a^5b^2$ **(b)** $18c^4d^5$ **(c)** $8p^3q^{12}$
     (d) $27d^9e^3$
4 $4n^2$
5 $64x^8y^6$

36. Expanding brackets

1 (a) $4y + 20$
     (b) $3y^2 - 3y + 6$
     (c) $5a + 10b - 15c$
2 $d^2 - 5d$

3 $6x + 4 + 5x = 11x + 4$

4 Both $= 6x + 6$

5 **(a)** $5y^2 - 30y$ **(b)** $4x^3 - 10x^2$ **(c)** $5y^3 + 5xy^2$

6 $y^2 + 5y - 14$

7 $4n^2 - 20n + 25$

37. Factorising

1 $4(t - 5)$

2 $3(3 + 7p)$

3 **(a)** $2a(a - 4)$ **(b)** $(n + 5)(n + 6)$

4 $a = 5, b = 2$

5 **(a)** $5d(2d - 3e)$ **(b)** $2x(5x + 4y)$

6 $(x - 9)(x + 9)$

7 $(4x - 5)(4x + 5)$

8 $(7x + 1)(x - 2)$

38. Algebraic fractions

1 $\dfrac{3}{2a}$

2 $\dfrac{6x + 19}{x^2 + 7x + 12}$

3 $\dfrac{2 - 13x}{4x^2 + 9x + 2}$

4 $\dfrac{x^2 - 2x}{15}$

5 $\dfrac{3}{10}$

6 $\dfrac{x - 4}{x - 2}$

7 $\dfrac{3x - 8}{x - 2}$

39. Straight-line graphs

1

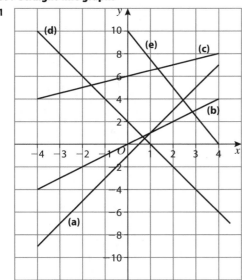

2 **(a)** $y = 2x + 2$ **(b)** $y = -\frac{3}{2}x - 1$

    (c) $y = -x$ **(d)** $y = 2x - 3$

40. Gradients and midpoints

1 $(3.5, 5.5)$

2 Both have gradient of $\frac{1}{2}$

3 $y = 2x + 8$

4 $y = \frac{5}{2}x - 4$ or $5x - 2y - 8 = 0$

41. Real-life graphs

1 **(a)** £15   **(b)** 0   **(c)** 15p

    (d) Plan A   **(e)** Cheaper by £9   **(f)** 460

42. Formulae

1 -3

2 18

3 3

4 **(a)** $C = 70d + 50$

    (b)

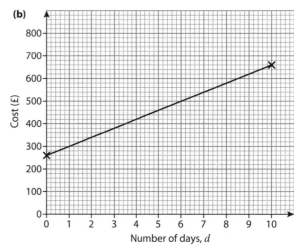

    (c) 8 days

43. Linear equations 1

1 **(a)** $x = 2$ **(b)** $x = 9$ **(c)** $x = 5$

    (d) $x = 7$ **(e)** $x = 4$ **(f)** $e = -10$

2 **(a)** $x = 4$ **(b)** $x = 5$

44. Linear equations 2

1 $x = 3$

2 $x = -7$

3 $x = 3$

4 $x = 3$

5 $x = 2\frac{1}{9}$

45. Number machines

1 **(a)** 11   **(b)** 6   **(c)** $2x + 5$

2 -1.5

3 -8

46. Inequalities

1 **(a)** $x > 0$ **(b)** $x \geqslant 7$

    (c) $x \leqslant 2$ **(d)** $x < -2$ **(e)**

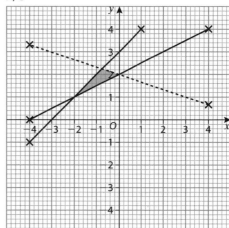

2 $-4, -3, -2, -1, 0, 1$

3 $n > 5$

4 $x \leqslant -2$

5 $x > 4$

6 $x < 1$

7 $n < -6$

47. Inequalities on graphs

1 A, C

2

48. Simultaneous equations 1

1 $a = 3, b = 1$

2 $x = -1, y = 7$

3 $x = -2, y = 4$

4 CD $=$ £8, DVD $=$ £12

49. Quadratic equations

1 $x = 3$ or -2

2 $x = 0$ or 5

3 $x = \pm 6$

4 $x = 0$ or 7

5 $x = \frac{1}{3}$ or -5

6 $x = 2$ or -5

50. Quadratics and fractions

1 (a) $3x^2 + 15x + 4x = 6x + 30, 3x^2 + 13x - 30 = 0$

  (b) $x = \frac{5}{3}$ or -6

2 $x = 0$ or 1

3 $x = 2$ or $-\frac{3}{2}$

4 $x = 2$ or -1

51. Simultaneous equations 2

1 $x = 6$ and $y = 3$, or $x = 1$ and $y = -2$

2 $x = 1$ and $y = 3$, or $x = -3$ and $y = -1$

3 $x = 3$ and $y = 4$, or $x = 4$ and $y = 3$

52. Rearranging formulae

1 $x = \dfrac{y - 1}{5}$

2 $b = 3(a - 2)$

3 $t = \frac{5}{2}p + 2$

4 $r = s(p - q)$

5 $x = 7y$

6 $a = \dfrac{2c + b}{3 - c}$

7 $x = \dfrac{4y - 5}{y - 1}$

53. Sequences 1

1 $3n + 4$

2 (a) $5n - 1$ (b) No, does not end in a 4 or a 9

3 (a) No   (b) 75

4 (a) No   (b) 11th term $= 118$

54. Sequences 2

1 $-2, 0, 2$

2 $0, 3, 8$

3 $n = 12$

4 $c = 3(3a - 4) - 4, c = 9a - 16$

5 3

55. Algebraic proof

1 $3 \times 2 + 3 = 9$, odd

2 $(2n + 1)^2 = 4n^2 + 6n + 1$
$= 2n(2n + 3) + 1$
'$2n$' = even, '$2n + 3$' = odd, even $\times$ odd = even, and even '+ 1' = odd

3 $2n(n + 1) + 1$
'$2n$' = even, '$n + 1$' = odd or even
even $\times$ odd = even, and even $\times$ even = even, and even '+ 1' = odd

4 $4n + 6 = 2(2n + 3)$

5 Sum: $n + (n + 4) = 2n + 4$
Difference of squares: $(n + 4)^2 - n^2 = 8n + 16 = 4(2n + 4)$
Difference of squares $= 4 \times$ sum

6 Sum: $(2n - 1) + (2n - 11) = 4n - 12$
Difference of squares: $(2n - 1)^2 - (2n - 11)^2 = 40n - 120$
$= 10(4n - 12)$
Difference of squares $= 10 \times$ sum

56. Identities

1 $n^2 + 4n + 4 + n^2 - 4n + 4 = 2n^2 + 8 = 2(n^2 + 4)$

2 $5x^2 + 10x + 5 - 5x - 5 = 5x^2 + 5x = 5x(x + 1)$

3 $a = -8, b = -18$

4 $a = -18, b = 99$

5 $a = 2, b = 5$

57. Completing the square

1 $(x + 2)^2 - 4; a = 2, b = -4$

2 $(x - 3)^2 - 10; a = -3, b = -10$

3 (a) $(x - 3)^2 - 6; p = -3, q = -6$

  (b) $x = 3 \pm \sqrt{6}$

4 $x = 4 \pm \sqrt{6}$

5 $x = 5 \pm 2\sqrt{3}$

58. Problem-solving practice 1

1 30, 60, 90   **2** £2.65   **3** $k = 8$

59. Problem-solving practice 2

4 $C = (0, 90), D = (60, 0)$

5 $n + (n + 1) + (n + 2) + (n + 3) = 4n + 6 = 2(2n + 3)$, 2 is even and as even $\times$ odd = even and even $\times$ even = even, so the four consecutive integers $[2(2n + 3)]$ are even

6 e.g. $n + 2 - n = 2, (n + 2)^2 - n^2 = 4n + 4, 2(n + 2 + n) = 4n + 4$

Unit 3

60. Proportion

1 60 seconds   **2** 96 minutes   **3** 45 litres

4 £14   **5** 9 hours

61. Trial and improvement

1 3.4   **2** 7.2   **3** 6.3

62. The quadratic formula

1 0.65 and -4.65 **2** 2.73 and -0.73

3 0.82 and -1.22 **4** 1.10 and -2.43

63. Quadratic graphs

1 (a)

x	-2	-1	0	1	2
y	3	-3	-5	-3	**3**

(b) 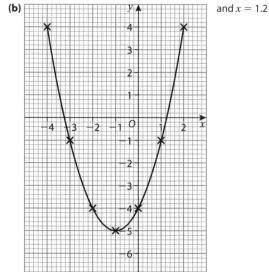   (c) -0.5

2 (a)

x	-4	-3	-2	-1	0	1	2
y	**4**	-1	-4	-5	-4	-1	4

(b) 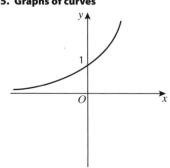   and $x = 1.2$

(c) $x = -3.2$ and 1.2

64. Using quadratic graphs

1 Draw $y = x - 1$ to give 4.3 and 0.7

2 Draw $y = 2x + 4$ to give 2.2 and -0.2

65. Graphs of curves

1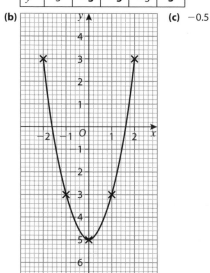

2

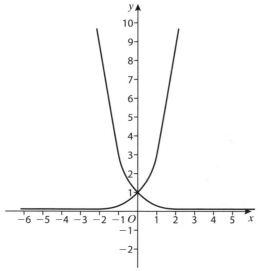

3 (a) B   (b) C   (c) A

66. 3-D coordinates

1 (a) $(3, 4, 0)$ (b) $(3, 4, 2)$ (c) $(1.5, 0, 1)$

2 (a) $(5, 3, 0)$ (b) $(2.5, 1.5, 0.5)$

67. Proportionality formulae

1 (a) $y = \dfrac{20}{x^2}$ (b) $\dfrac{5}{36}$

2 7.5

3 3

68. Transformations 1

1 (a) $y = 2x^2 + 3$

   (b)

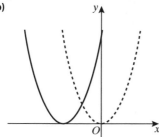

2 (a)

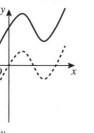

  (b)

  (c)

  (d)

69. Transformations 2

1 (a)

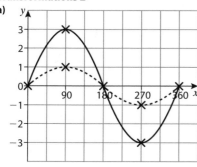

(b)

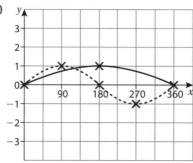

(c)

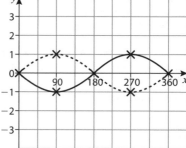

(d)

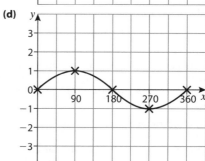

(e)

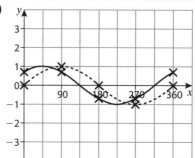

(f)

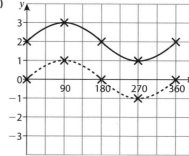

70. Angle properties

1 (a) 155°, corresponding angles are equal.
   (b) 155°, opposite angles are equal.

2 142°, allied angles add to 180°.

3 $4x - 20° = 180°$; $x = 50°$

71. Solving angle problems

1 70°; as MQN = 55° (angles on a straight line = 180°), so M = 70° (MQN is isosceles) and this is the opposite angle.

2 (a) 110°, alternate angles are equal.
   (b) 50°, angles in a triangle add to 180°.
   (c) 50°, opposite angles are equal.
   (d) 50°, alternate angles are equal.

72. Angles in polygons

1 exterior angle = $\frac{360}{6}$ = 60°, so interior angle = 180 − 60 = 120° (angles on a straight line = 180°)

2 **(a)** 72°   **(b)** 54°

3 10

73. Circle facts

1 **(a)** 55°   **(b)** 70°

2 28°

74. Circle theorems

1 **(a)** 43°   **(b)** 31°   **(c)** 53°
   (d) 70°   **(e)** 160°   **(f)** 101°

2 Angle *PRS* = 68° (Alternate segment theory)
Angle *SPR* = 32° (Angles in a triangle add up to 180°)
Angle *PQR* = 100° (Opposite angles in a cyclic quadrilateral add up to 180°)
Angle *PRQ* = 32° (Angles in a triangle add up to 180°)
So Angle *SPR* = Angle *PRQ*. These are alternate angles, so *PS* is parallel to *QR*.

75. Perimeter and area

1 **(a)** 24 m   **(b)** 32 m²

2 30 bags

3 275 cm²

76. Similar shapes 1

1 3.6 cm

2 **(a)** 7.5 cm
   (b) Vertical side: C is 3.75 times longer than A.
     Horizontal side: C is 3.66… times longer than A. Alternatively, show that the 11 cm side would be 11.25 cm if similar, or show that the 15 cm side would be 14.66… cm if similar.

3 6.5 cm

77. Congruent triangle proof

1 **(a)** SSS   **(b)** SAS   **(c)** AAS

2 AB = CD and BC = AD (opposite sides are equal in a parallelogram). AC is common, so triangle ABC is congruent to triangle ACD because SSS.

3 ST = SV and SP = SR (two sides of the same square).
Angle PST = RSV (SP and SR rotated the same as sides of the same square).
Triangle PST is congruent to triangle SRV (SAS), so angle PTS = angle RVS.

78. Pythagoras' theorem

1 43.01162… cm

2 37 inches

3 2101 m

79. Pythagoras in 3-D

1 12.68857… cm

2 22.71563… cm

3 18.70828… cm

80. Trigonometry 1

1 24.62431…°

2 65.2°

3 20.55604…°

81. Trigonometry 2

1 17.20729… cm

2 5.95876… km

3 9.94992… cm² or 10 cm²

82. The sine rule

1 25.37399…°

2 6.69244…cm

3 84.72534…m

83. The cosine rule

1 37.74299… cm

2 33.28641…°

3 48.62639…°

84. Trigonometry in 3-D

1 12.08977… m

2 13.16782… °

85. Circles and cylinders

1 16π cm²

2 πr + 2r

3 1847.3 cm³

86. Sectors of circles

1 **(a)** 32.9 cm   **(b)** 16π cm²

2 270°

87. Triangles and segments

1 £26 728.64

2 272.50190… cm²

3 54.35164… cm²

88. Prisms

1 **(a)** 146 250 cm³   **(b)** 25 650 cm²

2 35 cans of paint

3 54 cm

89. Volumes of 3-D shapes

1 64 cm

2 400 m³

3 **(a)** 15 spheres   **(b)** 476.401… cm³

90. Surface area

1 96π cm² or 301.59289… cm²

2 0.78π or 2.45044… m²

3 10.4 cm

91. Plans and elevations

1

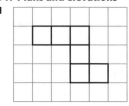

2

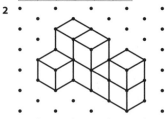

92. Bearings

1 **(a)** 040°   **(b)** 305°   **(c)** 120°   **(d)** 240°

2 **(a)** 050°   **(b)** 230°   **(c)** 40°

93. Scale drawings and maps

1 3 km

2 650 m

3 **(a)**

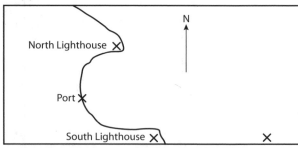

Scale: 1 cm to 10 km

   (b) South Lighthouse by 1 km

94. Constructions

1 Accurate drawing of triangle shown in Q1.

2 **(a)**

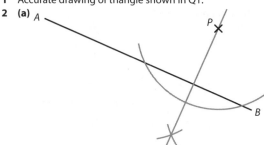

(b)

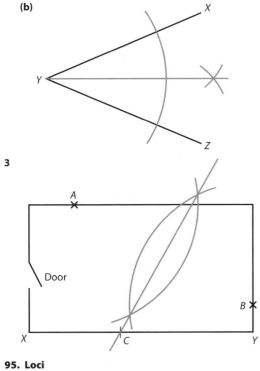

3

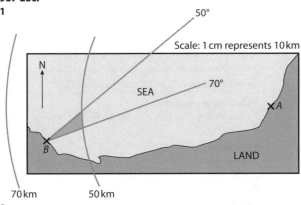

95. Loci

1

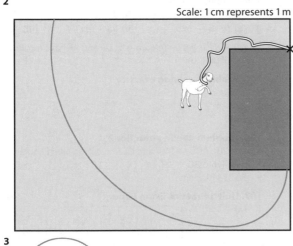

Scale: 1 cm represents 10 km

2

Scale: 1 cm represents 1 m

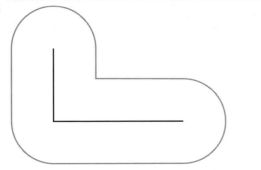

3

96. Speed
1 16 mph
2 3325 miles

3 3 seconds
4 24 mph
5 2.5 minutes
6 No, he arrives at 9.06 am.

97. Density
1 450 grams
2 210 grams
3 0.45 grams/cm^3
4 £957.28

98. Converting units
1 Yes, 104 km/h
2 30 m/s
3 **(a)** 157.5 litres   **(b)** 35 gallons   **(c)** 40 mpg
4 Almost, should be 3.75 m

99. Translations, reflections and rotations
1

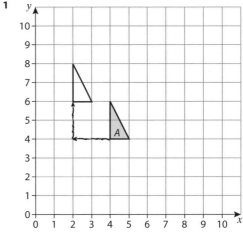

2 Rotation 90°, clockwise, about (4,3)
3 **(a)**

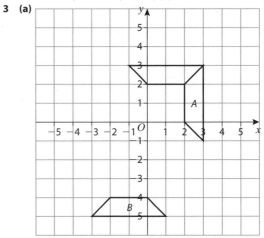

(b) Rotation 90°, anticlockwise, about (−2, 0)

100. Enlargements
1

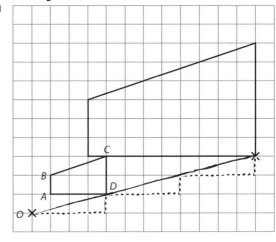

2

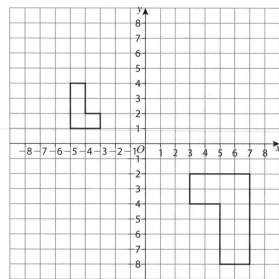

101. Combining transformations

1 (a)

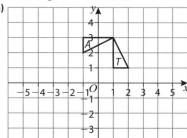

(b)

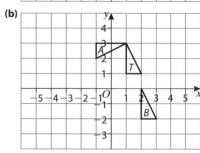

(c) Translation by the vector $\begin{pmatrix} -1 \\ 3 \end{pmatrix}$

2 (a)

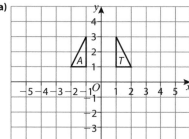

(b)

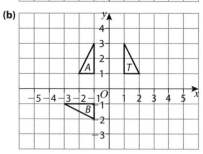

(c) Reflection in the line $y = -x$

3 (a)

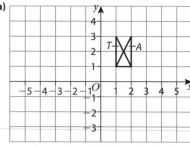

(b)

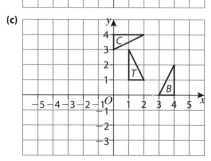

(c)

(d) Rotation 90°, anticlockwise, about (2, 3)

102. Similar shapes 2

1 (a) 1600 g   **(b)** 63.49604… mm

2 7.5 cm

103. Vectors

1 $\frac{1}{4}\mathbf{a} + \frac{3}{4}\mathbf{b}$

2 (a) 2**s**   **(b)** 3**t**

3 $\mathbf{b} - \mathbf{a}$

104. Solving vector problems

1 (a) $\mathbf{b} + \frac{1}{2}\mathbf{a}$ **(b)** $\mathbf{a} + \frac{2}{3}\mathbf{b}$ **(c)** $\frac{1}{3}\mathbf{b} - \frac{1}{2}\mathbf{a}$

2 (a) (i) $\mathbf{b} - 2\mathbf{a}$ **(ii)** $\frac{1}{3}\mathbf{a} + \frac{1}{3}\mathbf{b}$ **(iii)** $\frac{2}{3}\mathbf{b} + \frac{2}{3}\mathbf{a}$

  (b) $ADEF$ is a parallelogram as $\overrightarrow{DE}$ and $\overrightarrow{AF}$ have a common factor of $(\mathbf{a} + \mathbf{b})$.

105. Problem-solving practice 1

1 12 sides

2 2.55 cm

3 £600

106. Problem-solving practice 2

4 $x = 20°$, ADC = 140° and DCB = 40°, so allied angles, so AD is parallel to BC.

5 50.33340… cm²

107. Unit 1 Practice Exam Paper

1 (a)

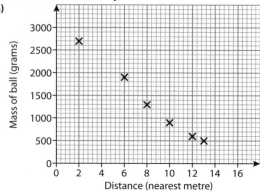

(b) Negative

(c)

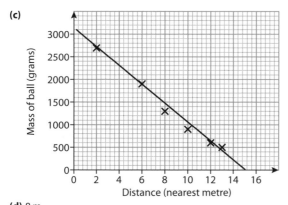

(d) 8 m

2 $2\frac{5}{6}$

3 £52.80

4 (a) £150   (b) Yes, £30

5 15

6 (a) $40 < t \leqslant 50$ (b) 51.2 seconds
(c) Data is grouped, so don't know exact values.

7 Yes, £2600 × 1.25 = £3250

8 (a)

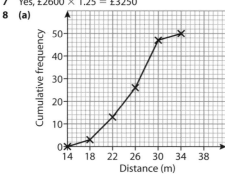

(b) (i) 25.6 m   (ii) 6.4 m

(c)

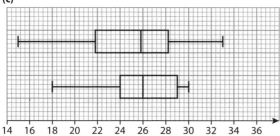

(d) Rounders ball – the distance thrown is more consistent than a cricket ball and the median is slightly further than a cricket ball. Alternatively, Cricket ball – it has been thrown 3 m further than a rounders ball, even though the lower quartile, median and upper quartile is less than with a rounders ball.

9 2, 9, 7, 6, 6

10 (a) 0.162   (b) 0.504

11 (a) 47   (b) £3880

12 0.125

112. Unit 2 Practice Exam Paper

1 20

2 (a) 16   (b) $2(n + 3)$

3 No, £20 cheaper.

4 (a) 70, 71, 72, 73, 74
(b)

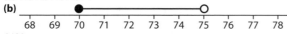

5 £480

6 (a) £315   (b) 2500 kWhr

7 (a) $x = 3.5$ (b) $4y + 13$ (c) $x = 3, y = 2$

8 (a) $4n + 2 = 2(n + 1)$, 2 is even.
If $n + 1$ is even, then $2(n + 1)$ is even × even = even.
If $n + 1$ is odd, then $2(n + 1)$ is even × odd = even.
(b) 26 (37th term to the 62nd term inclusively)

9

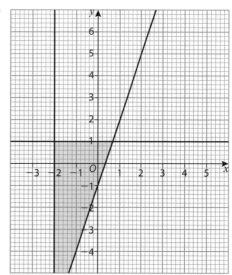

10 87%

11 $y = 2x - 7$

12 $\dfrac{x + 3}{2x + 5}$

13 (a) x^4 (b) $16a^{12}b^8$

(c) No, $125^{-\frac{1}{3}} = \dfrac{1}{\sqrt[3]{125}} = \dfrac{1}{5}$

14 (a) $(x - 4)^2 - 8$ (b) $4 \pm \sqrt{8}$

15 $37 + 10\sqrt{5}$

117. Unit 3 Practice Exam Paper

1 £45

2 $a = 36°$

3 $x = 130°$ (angles on a straight line = 180°),
$y = 130°$ (corresponding angles are equal)

4 59 cm²

5 (a) $T = 4x + 5$ (b) £18

6 (a)   (b) 2.1 m, 337°

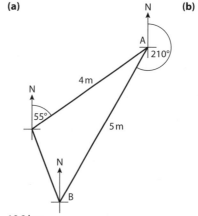

7 19.2 km

8 (a)

x	-3	-2	-1	0	1	2	3
y	10	**5**	2	1	**2**	5	**10**

(b)   (c) $y = 7.3$

9 (a) (i) 3   **(ii)** (1, 2)
(b)

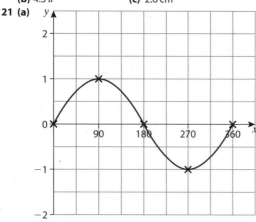

10

11 $x = 3.8\,\text{cm}$
12 $x = -7$
13 (a) $12.67854\ldots\,\text{cm}$ **(b)** $51.73595\ldots°$
14 $\dfrac{1.8}{3} = \dfrac{33}{55}$
15 (a) $50°$ **(b)** $40°$ **(c)** $130°$
16 $n^2 - 2n - 35 = 0, n = -5$
17 $OBD = \dfrac{180 - 2x}{2} = 90 - x$, $OBC = 90°$,
    $DBC = OBC - OBD = 90 - (90 - x) = x$
18 16
19 (a)

(b) (i) a   **(ii)** $-$b
    (iii) b $-$ a   **(iv)** b $-$ 2a
20 (a) $4\pi r^2 = 9\pi, r^2 = 2.25, r = 1.5$
    (b) 4.5π **(c)** $2.6\,\text{cm}$
21 (a)

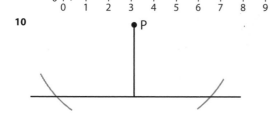

(b)

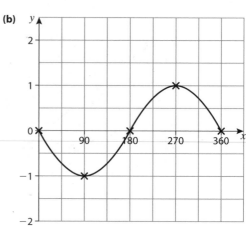

(c)

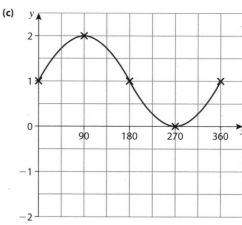

Published by Pearson Education Limited, Edinburgh Gate, Harlow, Essex, CM20 2JE.

www.pearsonschoolsandfecolleges.co.uk

Text and original illustrations © Pearson Education Limited 2013
Edited and produced by Wearset, Boldon, Tyne and Wear
Typeset and illustrated by Tech-Set Ltd, Gateshead
Cover illustration by Miriam Sturdee

The right of Greg Byrd to be identified as author of this work has been asserted by him in accordance with the Copyright, Designs and Patents Act 1988.

First published 2013

16 15 14 13 12
10 9 8 7 6 5 4 3 2 1

British Library Cataloguing in Publication Data
A catalogue record for this book is available from the British Library

ISBN 978 1 447 94144 6

Printed in Slovakia by Neografia

Every effort has been made to contact copyright holders of material reproduced in this book. Any omissions will be rectified in subsequent printings if notice is given to the publishers.

In the writing of this book, no AQA examiners authored sections relevant to examination papers for which they have responsibility.